Research at NaMLab

Band 5

T0135445

Research at NaMLab

Band 5

Herausgeber:

Prof. Dr.-Ing. Thomas Mikolajick

Stefan Döring

Scanning Spreading Resistance Microscopy and its Application to Passive and Active Semiconductor Device Characterization

Logos Verlag Berlin

Research at NaMLab

Herausgegeben von
NaMLab gGmbH
Nöthnitzer Str. 64
D-01187 Dresden

Bibliografische Information der Deutschen Nationalbibliothek

Die Deutsche Nationalbibliothek verzeichnet diese Publikation in der
Deutschen Nationalbibliografie; detaillierte bibliografische Daten sind
im Internet über http://dnb.d-nb.de abrufbar.

ISBN 978-3-8325-4450-8
ISSN 2191-7167

Logos Verlag Berlin GmbH
Comeniushof, Gubener Str. 47,
10243 Berlin
Tel.:   +49 (0)30 / 42 85 10 90
Fax: +49 (0)30 / 42 85 10 92
http://www.logos-verlag.de

Technische Universität Dresden

Scanning Spreading Resistance Microscopy and its Application to Passive and Active Semiconductor Device Characterization

Stefan Döring

von der Fakultät Elektrotechnik und Informationstechnik der
Technischen Universität Dresden

zur Erlangung des akademischen Grades eines

Doktoringenieurs

(Dr.-Ing.)

genehmigte Dissertation

Vorsitzender:	Prof. Dr. rer. nat. habil. Ehrenfried Zschech		
Gutachter:	Prof. Dr.-Ing. Thomas Mikolajick	Tag der Einreichung:	02.06.2016
	Prof. Dr.-Ing. Christian Boit	Tag der Verteidigung:	11.01.2017

Abstract

The aim of this work is the further development of the scanning spreading resistance microscopy (SSRM) technique. This scanning probe microscopy (SPM) method combines outstanding spatial resolution and a high dynamic range of the obtained electrical signal. Therefore, it is an excellent candidate for two dimensional characterization of semiconductor dopant distribution. The dopant-, or more precisely, active dopant or charge carrier distribution defines the functionality of every semiconductor device and therefore, its understanding is of major importance in semiconductor manufacturing.

After a short introduction and a review of alternative one- and two-dimensional dopant profiling techniques, in chapter 3 the key features of the SSRM technique are discussed, including the physics of the nano-electric point contact, sample preparation, the employed probes and data calibration. As a focus of this work was the applicability of SSRM to semiconductor failure analysis (FA), different site specific cross section preparation techniques are discussed. Thereby, improvements with respect to sample preparation and the associated effects on measurement data are shown. The dependency of the obtained SSRM signal from the polarity of the applied dc voltage between sample and probe tip is demonstrated. In many cases the obtained differences can be used to distinguish between p- and n-type doping, delivering important information for device characterization.

In the following chapter 4, the SSRM technique is employed at several application cases. This work features some of the few application examples to photovoltaic devices. Due to the focus on semiconductor FA, in the second section of this chapter two possible FA applications are presented. Especially the "Gate Depletion" example showcases the advantages of the analysis method: the possibility of precise, yet specific site preparation, free to choose in terms of position and orientation, an outstanding lateral resolution, well below 10 nm and the high dynamic range of the obtained spreading resistance signal, covering five orders of magnitude in free carrier concentration. In contrast to other SSRM studies, focusing on the application to advanced technology nodes of microprocessor units (MPUs) and the associated ever shrinking feature sizes, this work emphasizes the capability to obtain local two-dimensional dopant profiles not accessible by reference methods like secondary ion mass spectroscopy (SIMS) or spreading resistance profiling (SRP), not necessarily at the spatial resolution limit of SSRM. Therefore, the benefits for power device characterization are shown in a third application section. Moreover, the second power device is silicon carbide (SiC) based, a different semiconductor material, demanding special requirements for successful SSRM imaging.

Typically, SSRM measurements are performed at devices in equilibrium condition, as the individual device components are short circuited by the back contact. The fifth part of this work investigates the application of SSRM to actively driven devices. Beside the standard SSRM, the known scanning differential spreading resistance microscopy (SDSRM) and scanning voltage microscopy (SVM) are employed at a test device.

Additionally, a new method, combining SVM and SSRM in a two pass process is introduced within this work. The new scanning dynamic voltage spreading resistance microscopy (SDVSRM) method acquires the local sample surface voltage (V_s), dependent on the dc biases applied to the components of the actively driven device (V_{active}) in the first pass. The second pass consists of the actual SSRM measurement, in which the afore measured local surface potential is taken as input in order to dynamically adjust the terminal voltages in a way, that the local potential difference across the spreading resistance nano-electrical contact matches the software setpoint (V_{dc-sw}), while maintaining the internal voltage differences at the respective terminals.

Comparison of all methods applied to the test structure shows the newly developed SDVSRM to be the most accurate. SDVSRM adds a higher level of flexibility to SSRM, as occurring differences in cross section surface voltage are taken into account. These differences are immanent for actively driven devices, but can also be present at standard, short circuited samples. Therefore, the SDVSRM method offers the potential to improve characterization of samples at equilibrium conditions, too.

Contents

Acronyms 1
Symbols 4

1 Introduction   **7**
  1.1 The Semiconductor Industry 7
  1.2 Dopant Profiling Requirements 10

2 Review of Dopant Profiling Techniques   **13**
  2.1 1D Dopant Profiling Techniques 15
     2.1.1 Spreading Resistance Profiling 15
     2.1.2 Secondary Ion Mass Spectroscopy 20
     2.1.3 Electrochemical C-V Profiling 23
  2.2 2D Dopant Mapping Techniques 25
     2.2.1 Chemical Staining/Etching 25
     2.2.2 Secondary Electron Dopant Contrast 26
     2.2.3 Scanning Probe Techniques in General 27
     2.2.4 Scanning Capacitance Microscopy 30
  2.3 Other Dopant Analysis Techniques 33

3 SSRM Methodology   **35**
  3.1 SSRM as a Secondary AFM Imaging Mode 35
     3.1.1 Measurement Setup 37
     3.1.2 Nano-Electric Point Contact 40
     3.1.3 Back Contact 48
     3.1.4 Logarithmic Current Amplifier 51
  3.2 Sample Preparation 52
     3.2.1 Cleaving 54
     3.2.2 Mechanical Grinding and Polishing 56
     3.2.3 Focused Ion Beam 59
  3.3 Probes 64
     3.3.1 Diamond coated Probes 64
     3.3.2 Pyramidal moulded Diamond Probes 66
  3.4 Data Calibration 66

4 Application of SSRM for Technology Development   73
4.1 Photovoltaic 73
    4.1.1 Laser Doped Selective Emitter 74
    4.1.2 Other Selective Emitter Concepts 78
4.2 Integrated Silicon Devices 79
    4.2.1 Resist Lift Off 79
    4.2.2 Gate Depletion 84
4.3 Power Devices 100
    4.3.1 Super Junction MOSFET 101
    4.3.2 Silicon Carbide Power Devices 106

5 Active Device Characterization by SSRM   113
5.1 Motivation 114
    5.1.1 Dynamic Voltage 114
    5.1.2 Active Devices 115
5.2 Test Devices 117
    5.2.1 Sample Preparation 118
    5.2.2 Influence of Preparation on IV-characteristics 120
5.3 Existing Methods for Active Device Characterization 122
    5.3.1 Standard SSRM 122
    5.3.2 Lock-in Amplifier based SDSRM 126
    5.3.3 Scanning Voltage Microscopy 133
5.4 Scanning Dynamic Voltage Spreading Resistance Microscopy 138
    5.4.1 Measurement Principle 138
    5.4.2 Experimental Setup 140
    5.4.3 Results 144
    5.4.4 Summary 148

6 Summary and Outlook   149
6.1 General Summary 149
6.2 Outlook 150

Bibliography   153

Scientific contributions   171

Acronyms

4QPD	four quadrant photo diode
AFM	atomic force microscopy
AFP	atomic force prober
BEOL	back end of line
BL	bit line
BSF	back surface field
c-AFM	conductive AFM
CFM	chemical force microscopy
CG	gate capacitance
CMOS	complementary metal oxide semiconductor
CMP	chemical mechanical polishing
COX	gate oxide capacitance
CS	silicon channel capacitance
DAQ	data acquisition
DUT	device under test
ECV	electrochemical capacitance-voltage profiling
EOF	end of frame
EOL	end of line
FA	failure analysis
FEOL	front end of line
FET	field effect transistor
FIB	focused ion beam
FWHM	full width half maximum
GaN	gallium nitride
GIS	gas injection system
GOX	gate oxide
HF	hydroflouric acid
HNO_3	nitric acid
HSE	height selective emitter

IC	integrated circuit
IGBT	insulated gate bipolar transistor
ILD	inter layer dielectric
InP	indium phosphide
KPFM	kelvin probe force microscopy
LDSE	laser doped selective emitter
MEMS	micro electro-mechanical systems
MFM	magnetic force microscopy
MOS	metal oxide semiconductor
MOSFET	metal oxide semiconductor field effect transistor
MPU	microprocessor unit
OM	optical microscopy
PD	photodiode
PFA	physical failure analysis
PID controller	proportional-integral-derivative controller
poly-Si	polycrystalline silicon
PSE	printed selective emitter
PSL	polysterene latex
PV	photovoltaic
RMS	root mean square
ROM	read only memory
SAM	signal access module
SCF	single cell fail
SCM	scanning capacitance microscopy
SCR	space charge region
SDSRM	scanning differential spreading resistance microscopy
SDVSRM	scanning dynamic voltage spreading resistance microscopy
SEDC	secondary electron dopant contrast
SEM	secondary electron microscopy

SiC	silicon carbide
SIMS	secondary ion mass spectroscopy
SJ	super-junction
SMM	scanning microwave microscopy
SMU	source measure unit
SNOM	scanning near field optical microscopy
SPM	scanning probe microscopy
SRAM	static random access memory
SRP	spreading resistance profiling
SSRM	scanning spreading resistance microscopy
SThM	scanning thermal microscopy
STI	shallow trench isolation
STM	scanning tunneling microscope
SVM	scanning voltage microscopy
TCAD	technology computer aided design
TEM	transmission electron microscopy
ToF-SIMS	time of flight secondary ion mass spectroscopy
TTL	transistor transistor logic
TUNA	tunneling AFM
UHV	ultra high vacuum
USB	universal serial bus
VC	voltage contrast
WL	word line

Symbols

E_{crit}	critical electrical field strength - $[V/m]$
ε_0	permittivity of free space $\varepsilon_0 = 8.854188 \cdot 10^{-12}\,\mathrm{F/m}$
ε_{LJ}	depth of the potential well of the system exhibiting a Lennard-Jones potential - $[eV]$
ε_r	relative permittivity of the material
I_{ON}	on state current - $[A]$
I_t	tunneling current - $[A]$
μ	carrier mobility - $[cm^2/V \cdot s]$
μ_n	mobility of electrons in a specific material - $[cm^2/V \cdot s]$
μ_p	mobility of holes in a specific material - $[cm^2/V \cdot s]$
N	amount of dopant atoms in silicon crystal lattice per cm^3.
N_A	amount of acceptor atoms in silicon crystal lattice per cm^3.
N_D	amount of donor atoms in silicon crystal lattice per cm^3.
n_i	intrinsic carrier concentration in the crystal material - $[1/cm^3]$
q	elementary charge; the absolute value of the charge carried by a singe electron $q = 1.602177 \cdot 10^{-19}\,\mathrm{C}$
ρ	specific resistance of a material. $\rho = R \cdot \frac{A}{l}$ - $[\Omega \cdot m]$
ρ_{LJ}	characteristic distance of the Lennard-Jones potential, where attractive and repulsive forces cancel each other out - $[m]$
R_{meas}	measurement resistance, including all resistance contributions and the software processing - $[\Omega]$
R_{ON}	on state resistance - $[\Omega]$
σ	electrical conductivity - $[S/m]$
U_{LJ}	Lennard-Jones potential - $[eV]$
V_{Acc}	acceleration voltage - $[V]$

V_{BR}	breakdown or blocking voltage of a device - $[V]$
V_{active}	dc bias applied to one component of the actively driven device - $[V]$
V_{dc}	dc sample bias applied between sample and probe - $[V]$
V_{dc-hw}	dc sample bias applied to the chuck - $[V]$
V_{dc-sw}	dc sample bias as set in the AFM controller software and used for resistance calculation - $[V]$
$V_{dynamic}$	dynamic dc bias applied to one component of the actively driven device - $[V]$
V_{probe}	SPM probe tip voltage; in SSRM mode, $V_{probe} := 0\,\mathrm{V}$ - $[V]$
V_s	local effective sample surface voltage - $[V]$
v_{sat}	electron saturation velocity - $[^{cm}/_s]$
V_{Th}	threshold voltage - $[V]$

1 Introduction

In this chapter a motivation is given for the development and the application of scanning spreading resistance microscopy (SSRM), a two-dimensional imaging technique of free carriers, respective dopants with high spatial resolution and a high dynamic range. It is shown that both, development and applications, were, and still are driven by the semiconductor industry. A brief history of the semiconductor industry is given. It is shown, that SSRM is beneficial in every phase of a semiconductor product cycle, starting with the early technology development until failure analysis of field returns.

The second part of the introduction gives an overview of other established techniques for doping and carrier profiling. Advantages and limitations of the particular methods in comparison to SSRM are discussed with respect to spatial resolution, dynamic range and sensitivity.

1.1 The Semiconductor Industry

The origins of the semiconductor industry can be traced back to the nineteenth century. As early as 1883 selenium rectifiers have been built on an experimental basis [1]. The first commercially important use of semiconductors dates back to 1904. At this time the materials ability to detect high frequency currents was discovered [2]. As the demand for the thermionic valves grew rapidly, due to its application in radar, telephone technology and broadcast, an effective replacement was necessary and the semiconductor research gained momentum. In 1945 the research group "semiconductor" was found at Bell Laboratories under the lead of William Shockley. Their goal was to develop a field effect transistor (FET) as described by Lilienfeld in 1925 [3, 4].

In 1947 Bardeen, Brattain and Shockley invented the point-contact transistor at Bell Laboratories [5, 6]. The famous experimental setup is shown in figure 1.1. It consists of a block of germanium and two gold contacts very close to each other. The gold contacts are attached to a plastic triangle. Schottky contacts are formed by pushing the triangle and the attached gold contacts with a spring against the germanium surface. Although the device was crudely constructed, it was a break through.

Figure 1.1: *Experimental setup of the first transistor. (source: Bell Labs)*

Another milestone was the announcement of the first silicon transistor by Gordon Teal from Texas Instruments in 1954 [7] [1]. The next major step followed in 1960, when Jack Robert Noyce invented the first monolithic integrated circuit.

The potential of integrated semiconductor devices was first identified by Gordon Moore [8]. He forecasted a doubling in chip complexity every year. Although Moore's law, as his outlook was dubbed, was adjusted slightly later on, the conclusion remains the same: the cost per transistor decreases, while device complexity increases (figure 1.2). Driven by Moore's law, semiconductor products replaced a lot of other components as the afore mentioned thermionic valves. Moreover, the industries momentum created entirely new devices, applications and businesses.

In 2015 a single memory chip or microprocessor unit (MPU) includes billions of single transistors. The smallest feature size for integrated circuits (ICs) in production is as small as 14 nm. Beside the high end memory and MPU chips, silicon semiconductor devices are also used as power switches in all the different voltage and consumption classes, as micro electro-mechanical systems (MEMS) or sensors (e.g. photodiodes). Silicon solar cells are used for energy conversion.

---

[1]Although it turned out Tanenbaum created the first silicon transistor just a few months earlier at Bell Labs [7].

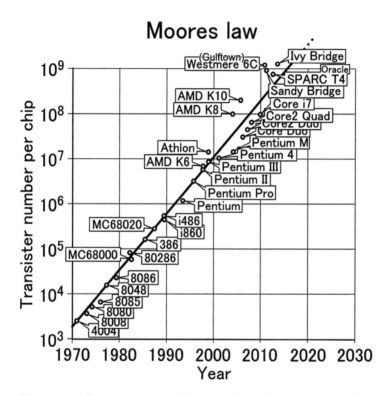

Figure 1.2: *Transistor counts of integrated circuits plotted against their dates of introduction. The data is in good agreement with Moore's law (plotted line) - the doubling of transistor count every two years.*

All of these applications require the analysis of the local dopant-, respectively free carrier concentration with high spatial resolution and a high dynamic range for a successful development, failure analysis and production support.

1.2 Dopant Profiling Requirements

The goals and the advancement of the semiconductor industry are coordinated by the ITRS, the International Technology Roadmap for Semiconductors [2]. The roadmap assesses the semiconductor industry´s future technology requirements for the next 15 years. One of the sub chapters of the organization is dedicated to metrology. Here the requirements for dopant-/carrier profiling can be found. Figure 1.3 shows an excerpt of the long term front end metrology requirements taken from the 2001 edition of the ITRS. The expected junction depths are listed as well as the required lateral steepness of the dopant profile, the lateral/depth resolution for 2D and 3D profiling and the dopant profile concentration precision.

YEAR OF PRODUCTION	2010	2013	2016
DRAM ½ PITCH (nm)	45	32	22
MPU / ASIC ½ PITCH (nm)	45	32	22
MPU PRINTED GATE LENGTH (nm)	25	18	13
MPU PHYSICAL GATE LENGTH (nm)	18	13	9
Dopant concentration (atoms/cm³) Dopant atom	1.4×10^{19} P, As, B	2.0×10^{19} P, As, B	$? \times 10^{19}$ P, As, B
Metrology for junction depth of (nm)	7	5	4
Lateral Steepness of dopant profile (nm/decade)	2.7	1.9	1.6
Lateral/depth spatial resolution for 2D / 3D dopant profile (nm)	2.2 / 2.0	1.5 / 1.4	1.0 / 1.2
At-line dopant profile concentration precision (across concentration range) [E]	2%	2%	2%

White—*Manufacturable Solutions Exist, and Are Being Optimized*

Yellow—*Manufacturable Solutions are Known*

Red—*Manufacturable Solutions are NOT Known*

Figure 1.3: *ITRS 2001 Edition - Dopant-/Carrier Front End Processes Metrology Technology Requirements — Long-term*

For comparison the metrology table from the 2012 update is shown in figure 1.4. It turns out that the predictions of 2001 are met quite well. Only the physical gate length did not shrink as fast as predicted. The metrology requirements for dopant profiling relaxed a little over time as one can see comparing the 'Lateral Steepness of dopant profile' and 'Lateral/depth

---

[2]http://www.itrs2.net/

spatial resolution for 2D/3D dopant profile' values. As apparent in the tables, in 2001 there were no solutions known to meet the future demands with respect to dopant profiling. The situation changed until 2012, as in the new report metrology solutions exist or are contrivable at least. One of the analysis methods enabling the high resolution profiling is SSRM. However, these ITRS metrology requirements can be taken as a benchmark for 2D/3D dopant-/carrier profiling.

Year of Production	2011	2012	2013	2014	2015	2016
Flash ½ Pitch (nm) (un-contacted Poly) (f)	22	20	18	17	15	14,2
DRAM ½ Pitch (nm) (contacted)	36	32	28	25	23	20,0
MPU/ASIC Metal 1 (M1) ½ Pitch (nm) (contacted)	38	32	27	24	21	18,9
MPU Printed Gate Length (GLpr) (nm) ††	35	31	28	25	22	19,8
MPU Physical Gate Length (GLph) (nm)	24	22	20	18	17	15,3
Dopant atom	P, As, B	P, As, B	P, As, B	P, As, B	P, As, B	P, As, B
Drain extension X_j (nm) for bulk MPU/ASIC [A]	10,5	9,5	8,7	8	7,3	
Extension lateral abruptness for bulk MPU/ASIC (nm/decade) [C]	2,8	2,4	2,3	2	1,8	
Lateral/depth spatial resolution for 2D/3D dopant profile (nm)	2,8	2,4	2,3	2	1,8	
At-line dopant concentration precision (across concentration range) [D]	2%	2%	2%	2%	2%	2%

Figure 1.4: *ITRS 2012 Edition - Dopant-/Carrier Front End Processes Metrology Technology Requirements*

It has to be kept in mind that the requirements apply for metrology of the junction depth (2001) and the drain extension of a bulk MPU/ASIC (2012). Analyzing different devices or device features might place different, partly much more relaxed, partly harder requirements.

The need for two dimensional dopant profiling also results from the transition of former planar devices through to three dimensional devices. For the high end MPUs this trend started with the introduction of Fin-FETs [9]. However, this requirement is not limited to the advanced MPUs. Latest generation NAND-Flash also uses 3D integration [10]. The same trend towards 3D integrated devices is visible for the different power semiconductor devices.

Three major applications for dopant-/carrier profiling may be distinguished:

- process characterization

- device simulator matching

- failure analysis (FA)

In process characterization a device is analyzed with respect to one or more process steps. The device is not necessarily fully processed. Dopant area formation can be observed for example with respect to lithography related parameters (e.g. used mask, mask alignment or used resist), implant conditions (acceleration voltage (V_{Acc}), implant dose, implant species) or annealing conditions. If several annealing processes are applied to form a certain dopant area, process characterization could analyze the specific area in between the different steps to illustrate the progress of the diffusion. For fully processed devices, analysis data can also be cross correlated to electrical device characteristics.

As device simulator models represent approximations to the real world physics, it is necessary to verify the simulator results. Most simulators for example lack a diffusion model for complex (co)-implants, as germanium or carbon. Not limited to single isolated devices, the analysis carried out for device simulator matching allows obtaining information on device formation under different embedding conditions (isolated, nested, dense packaging). The findings can be used to adjust the simulation properly.

FA is an integral part of the semiconductor industry. It accompanies a product or technology from the early phases of development over the ramp up process and yield learning even after chip sales if customer returns occur. As production processes include hundreds or thousands of single process steps, there is a vast number of defect sources. The FA goal is to track down the defects root cause, enabling the application of corrective actions. Dopant-/carrier related failure analysis is mostly driven by front end of line processes. Common examples are misaligned implant masks, or masked implants.

2 Review of Dopant Profiling Techniques

This section gives a review of commonly used dopant profiling techniques. Popular one dimensional and two dimensional profiling methods are described.

The different analysis methods can be divided into dopant- or carrier sensitive. Dopant sensitive techniques analyze the different types of atoms representing the donors and acceptors incorporated in the semiconductor lattice. Moreover dopant related techniques offer the possibility to determine what kind of dopant species was used (e.g. phosphor or arsenic as donor atoms for n-type doping). However a more precise description of the device functionality and performance can be concluded from the knowledge of the actual carrier concentration distribution. Although dopant and carrier distribution are linked closely, they do not necessarily match. Possible differences can originate from three main sources. The first one can be illustrated at the determination of the p-n-junction position. Whereas the dopant sensitive techniques exhibit the metallurgic junction, carrier sensitive techniques show the electrical junction position (see figure 2.1). Only for symmetric junctions (N_A=N_D) the metallurgic junction coincides with the electrical p-n-junction.

The second possible deviation between dopant- and carrier sensitive techniques is rooted in the fact, that as implanted not all dopant atoms are activated and contribute to the free carrier distribution. Only for a complete activation, dopant and carrier concentration are equal. The third explanation for possible differences involves changes in the free carrier mobility μ. As the current density σ, which is accessed for example in resistance measurements, is the product of the carrier concentration N and their mobility μ, changes in local mobility directly influences measurement results. Although there are these differences, for the sake of simplicity, we will refer to the described techniques as dopant profiling techniques. In the following review of dopant profiling techniques, key figures for maximum depth- or spatial resolution and the dynamic range of the methods will be given. As seen in 1.2 the need for high depth- and spatial resolution is a direct deduction from the ITRS

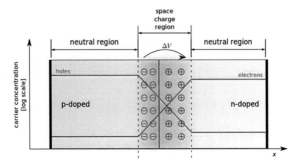

Figure 2.1: *Schematic drawing of the dopant- and carrier distribution at an asymmetric p-n-junction.*
(source: This vector image was created with Inkscape by Adundovi. (Own work) [CC BY 3.0 (http://creativecommons.org/licenses/by/3.0) or GFDL (http://www.gnu.org/copyleft/fdl.html)], via Wikimedia Commons)

metrology requirements. In general 1D-profiling and 2D-mapping analysis techniques and their results benefit from and therefor target for as high spatial resolution as possible. The maximum achievable dynamic range of a specific analysis method is also very important for dopant profiling and mapping. Common dopant densities within semiconductor devices span a range from as small concentrations as 10^{13} 1/cm³ up to 10^{21} 1/cm³ for example for the highly doped source drain implants. The ideal metrology method allows obtaining data over the whole range.

Another important factor is the techniques applicability to really integrated devices in contrast to the limitation to test structures. During the development process of a device or a technology, it might be sufficient to gain data from specific test structures. Usually these test structures are located in the kerf region between the individual chips on a wafer[1] or an appropriate test chip is build. However, failure analysis (FA) for yield learning or customer returns processing makes inevitable to be able to prepare and analyze not only test structures, but (in an ideal case) every single feature integrated in the chips. Of course reverse engineering has the same requirements.

---

[1]This area is reserved for the later saw cut that separates the chips.

2.1 1D Dopant Profiling Techniques

Three of the most common one dimensional dopant analyzing techniques should be noted here. All of them are well accepted and standard methods in semiconductor industry. The electrochemical $C - V$ profiling (2.1.3) is very wide spread in silicon based photovoltaic (PV).

2.1.1 Spreading Resistance Profiling

Spreading resistance profiling (SRP) is applied in semiconductor industry since the 1960s. It was proposed by Mazur and Dickey [11]. Although originally used to analyze lateral dopant variations in a top-down approach, today the main application of SRP is dopant depth profiling. The technique is the predecessor of scanning spreading resistance microscopy (SSRM) and exhibits many similar aspects. Therefore SRP is discussed here in more detail.

As it is an electrical measurement technique, it senses carrier concentration, rather than dopant concentration. The method offers a very high dynamic range of 10^{12} to 10^{21} $^{1}/_{cm^3}$ in terms of dopant concentration. Reproducibility of 10% can be achieved [12]. The maximum achievable depth resolution is in the range of a few nanometers. In general SRP is applied to test structures or blanket wafers. Putting some efforts into sample preparation and measurement setup, it becomes possible to extract dopant profiles from integrated devices with lateral homogeneity in the range of $10\,\mu m$. This range is defined by the measurement setup as we will see in the next paragraph.

In SRP two well aligned metallic probes with a small voltage difference are stepped over a beveled semiconductor cross section surface. A schematic of the measurement setup is shown in figure 2.2.

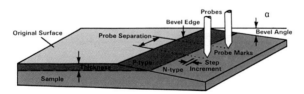

Figure 2.2: *Schematic drawing of the SRP setup. Bevel cross section is visible at the right hand side of the sample. Stepped probe path is indicated by the dotted lines. (source: SGS Institute Fresenius)*

At each step the resistance between the probes is measured [11, 13]. The measured resistance is given by the following equation:

$$R_{meas} = 2R_{probe} + 2R_{cont} + 2R_{spread} \qquad (2.1)$$

with R_{probe} being the probe resistance, R_{cont} the contact resistance and R_{spread} the spreading resistance which should be assessed. The principle of the method relies on the dominance of the spreading resistance compared to the other serial resistance contributions in the measurement path.

As with most of the direct probing techniques, sample preparation is important for successful measurement. Usually at the start of preparation, all metallization and inter layer dielectrics are removed. This can be done wet chemically by an hydroflouric acid (HF) lift off procedure. It is of advantage to cover the so exposed silicon surface with an insulator (oxide or nitride) in order to get a well-defined start of the subsequently created bevel cross section. The so prepared sample is attached to a plunger or piston that offers a predefined tilt. Bevel angles as small as 1° are achievable in a reproducible way by this method. The piston with the bonded sample is mechanically grinded and polished on a spinning glass plate with a diamond suspension on top. The glass plate surface must be very uniform with a peak-to peak roughness of approximately 130 nm. In order to achieve high quality bevel surfaces diamond suspensions with grain sizes down to 50 nm have to be used. Finished with polishing the realized bevel angle (especially the very small angles for the high resolution measurements) can be controlled by a profilometer or optically by the laser reflection from the sample surface and the bevel cross section onto a wall at a known distance. In general, the smaller the bevel angle, the higher the depth resolution. The step size (Δx) and the bevel angle (θ) define the equivalent depth (Δz):

$$\Delta z = \Delta x \cdot \frac{sin(\theta)}{cos(\theta)} \qquad (2.2)$$

For small angles equation 2.2 can be simplified as:

$$\Delta z = \Delta x \cdot sin(\theta) \qquad (2.3)$$

For example a step size of 5 μm applied to a surface with a bevel angle of 1° result in an equivalent step in z of 87 nm.

The SRP measurement is started some microns before the bevel cross section begins to determine the transition in the measurement data or to be able to count the probe imprints on the sample surface until the bevel cross section is reached. The two probes have to be well aligned parallel to the bevel. Deviations in the alignment result in decreasing z resolution.

To understand spreading resistance, a cylindrical metallic probe is considered contacting a flat semiconductor surface (figure 2.3). In our example a current I flows through the cylindric probe of diameter $2r$ into a semiconductor of resistivity ρ. After passing the transition from the probe to the semi-infinite semiconductor the current is not confined anymore and spreads out. Due to this current spreading, the method is called *Spreading Resistance Profiling* [2].

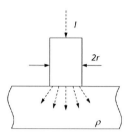

Figure 2.3: *A cylindrical metallic probe of the diameter $2r$ in contact with a flat semiconductor surface with a homogenous resistivity ρ. The dotted arrows indicate the current flow.*

It is shown [14] that for a non-indenting, cylindrical contact of a highly conductive probe with a planar, circular interface, the spreading resistance is

$$R_{spread} = \frac{\rho}{4r} \tag{2.4}$$

with the resistivity ρ

$$\rho = \frac{1}{q(n\mu_n + p\mu_p)} \tag{2.5}$$

In 2.5 q is the elementary charge, n and p the density of electrons and holes and μ_n and μ_p are the mobilities for electrons respectively holes contributing to the current flux.

Assuming a hemispherical, indenting probe tip of radius r, the spreading resistance becomes

---

[2] and Scanning *Spreading* Resistance Microscopy

$$R_{spread} = \frac{\rho}{2\pi r} \tag{2.6}$$

With equation 2.6, the measured resistance during an SRP measurement (equation 2.1) can be written as

$$R_{meas} = 2R_{probe} + 2R_{cont} + \frac{\rho}{2\pi r} \tag{2.7}$$

Typical resistivity values of silicon are in the range of $10^{-3}\,\Omega$cm for high dopant densities of $\approx 10^{20}\,1/\text{cm}^3$ and $100\,\Omega$cm for low dopant densities of $\approx 10^{14}\,1/\text{cm}^3$ [15, 16]. Since SRP uses probes of small tip radii around $1\,\mu$m equation 2.6 predicts $R_{spread} \approx 1600\rho$. It is the small tip radii used in SRP, enabling the spreading resistance part in the measurement path to become dominant in comparison to other contributors as probe resistance and contact resistance.

A closer look exhibits measurement data to be more complicated. The probe resistance R_{probe} can be considered constant. But the contact and spreading resistance R_{cont} and R_{spread} are influenced by several factors that have to be taken into account [17, 18].

$$R_{meas} = 2R_{probe} + 2R_{cont}(\rho) + \frac{\rho}{2\pi r(\rho)}CF[\rho, S, r(\rho)] \tag{2.8}$$

R_{cont} depends on the semiconductors resistivity, the density of surface states and, as described later, is strongly dependent on the materials phase beneath the probe tip. In the spreading resistance term a correction factor (CF) has to be introduced including resistivity, probe spacing (S) and contact radius r. The correction factor (or Schumann and Gardner correction factor [19]) takes the influence of the current distribution from the underlying resistivity profile into account. An underlying layer of higher resistivity would result in an increased measurement resistance compared to the bulk material resistance. Other way around, a buried more conducting layer leads to reduced measurement resistance values.

The situation gets even more complicated for samples including p-n junctions. Carrier spilling occurs at a bevel cross section through the junction [20]. It is a result of the material removal due to the beveling process and the associated redistribution of mobile carriers.

Although a hemispherical probe tip is assumed in 2.6, in reality the probes are conditioned before use to reproducibly form micro contacts. The presence of micro contacts is known since shortly after the method became popular [21, 22] and is believed to support the probes breakthrough of the native

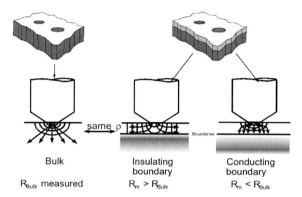

Figure 2.4: *Effect of underlying layers on electric field at spreading resistance conditions, showing that calibration curves apply only for a homogeneously doped bulk semiconductor. For underlying layers with higher resistivity the measured resistance increases and for layers of smaller resistivity compared to the sample surface layer, the resistance obtained in an SRP decreases. (source: SGS Institute Fresenius)*

oxide. The conditioning on purpose leads to an improved contact resistance and improved reproducibility of the SRP probe contact [23].

The influence of surface states has been discussed [24]. For highly damaged surfaces as the bevel cross sections prepared for SRP, a high amount of surface states can be concluded. The resulting Fermi level pinning is expected to result in most p-type materials exhibiting a depletion layer whereas n-type material will have an inversion layer at the surface.

An approximation considering a hemispherical probe tip of $1\,\mu$m diameter and a probe load of 5 g (which are commonly used values) results in an estimate of the contact pressure of approximately 16 GPa. Pressure values beneath the micro contacts can be expected to be significantly higher, whereas the flat contact areas in between would exhibit smaller pressure values. A phase transition in the silicon material is known to take place from the cubic Si(I) phase to the body centered tetragonal Si(II) phase ($\beta - Sn$) exceeding pressure values of about 11 GPa [25, 26]. Experiments exhibited the $\beta - Sn$ phase to have metallic material properties in contrast to the Si(I) phase. A resistance drop of approximately 5 orders of magnitude was observed when the transition to the Si(II) occurred [25]. The same phase transition takes place in the SRP measurement providing good contact resistance R_{cont} between the probe tip and the semiconductor surface.

SRP is a comparative technique. Calibration samples of known resistivity are measured for a specific set of probes at a specific time. The data acquired is used to generate calibration curves for the specific measurement setup. These calibration measurements have to be carried out on a regular basis in order to guarantee the comparability of device under test (DUT) measurement data. In general, these calibration curves can be applied in a straight forward way only for samples of homogenous resistivity. Existing resistivity profiles in depth make the aforementioned correction factor CF necessary.

The corrections and influences (correction factors, carrier spilling, surface damage, micro contact distribution and three dimensional current flow) scale with probe radius and probe separation and become increasingly important for shallow profiles. Probe penetration and bevel roughness decrease depth resolution further. All of those effects limit the applicability of SRP to layers smaller than 100 nm drastically.

The profiling in one dimension limits the range of application for SRP further. Therefore some efforts were made, to enhance the method for two dimensional data acquisition [27, 28]. At about the same time SSRM was developed, exhibiting superior imaging properties.

2.1.2 Secondary Ion Mass Spectroscopy

In secondary ion mass spectroscopy (SIMS) a primary ion beam is focused on the sample, sputtering surface atoms. A part of those sputtered ions is ionized and gets subsequently collected, mass separated and counted. Due to the sputtering of surface atoms (by the primary beam or in an additional step in between the measurements) deeper layers can be measured successively. As a result, depth profiling can be achieved.

First SIMS instruments were independently developed in the early 1960 by Castaing and Slodzian [29] and Herzog and Liebl [30]. It took 10 years until Benninghoven improved depth resolution and sensitivity drastically by lowering primary ion densities [31], so the method became interesting for (semiconductor) surface analysis.

In contrast to SRP measuring active carriers, SIMS delivers information on the chemical composition of the sample. Data from the same sample obtained with the two different methods can exhibit significant differences. The first difference relies on dopant activation. Analysis of an implanted but not activated sample using SIMS would show the Gaussian distributed total amount of impurity atoms. As most of the dopant atoms are not activated until an activation anneal, methods based on carrier analysis as SRP or SSRM would show a far weaker corresponding signal. For annealed samples, SIMS and electrical measurements agree quite well [32]. The second difference is based on possible differences between the dopant distribution and the carrier distribution in a non-homogeneously doped sample; especially at $p - n$ junctions (see figure 2.1).

Under ideal conditions SIMS can achieve a depth resolution of 1 to 5 nm and a dynamic range of 10^{14} to 10^{21} $1/\text{cm}^3$. Typical spot sizes for a SIMS measurement are in the range of $40 \, \mu\text{m}^2$. This area is too large for most of the FA requirements, as FA is often done at very localized defects. SIMS is usually applied at test structures or blanket wafers. It is commonly used as one dimensional depth profiling method.

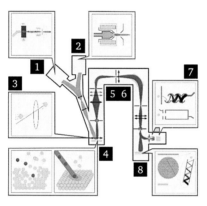

Figure 2.5: *basic SIMS setup showing the ion gun (1 or 2), generating the beam that is focused (3) on the sample. Surface atoms are ionized and sputtered (4). The secondary ions are collected and mass filtered (5, 6) before counting at an electron multiplier (7, top) a Faraday cup (7, bottom) or a CCD screen (8).*

The basic measurement setup is shown in figure 2.5. Primary ions are generated by an ion gun. Different ions can be used for analysis. Figure 2.5 schematically shows a surface ionization source (1 - e.g. cesium source generating $^{133}Cs^+$) and a duoplasmatron (2 - ions of gaseous elements as $^{40}Ar^+$). Liquid metal ion guns are also in use for ion generation (e.g. gallium sources). Within the ion column the primary ions are accelerated and focused on the sample. Optional filtering, pulsing and scanning of the primary beam are also done here.

When the primary ion impulse is high enough, sample atoms of the uppermost layer can be emitted (sputtered). Representative sputter yields are in the range of 1 to 5 surface atoms per primary ion. Based on the total amount of sputtered surface atoms, only a small amount (typically $< 5\%$) is ionized. These secondary ions are collected within the secondary ion column, providing guidance of as much secondary ions to the detector as possible. Prior detection, the secondary ions are mass filtered, whether by a quadrupole mass spectrometer or as indicated in figure 2.5 by a sector field mass spectrometer. The quadrupole type spectrometer offers higher switching speeds between the masses to be separated, allowing fast depth profiling while monitoring different sputtered species. On the other hand, the two combined filter mechanisms of a sector field mass spectrometer provide a higher mass resolution $(\frac{m}{\Delta m})$.

Now, that the beam consists only of particles the same species (rather the same mass), these particles are detected. Different types of detectors are possible, as shown in figure 2.5. The analysis time can be translated into analysis depth by measuring the total sputter crater depth after SIMS analysis and knowing the total analysis time. In the following signal processing, the detected secondary ion count is plotted over analysis depth. When switching the mass spectrometer during the SIMS analysis for filtering different masses in quick succession, depth profiles of several elements can be obtained simultaneously.

Nevertheless, SIMS is a comparative method. Samples of known composition are characterized using the exact same measurement conditions as with the DUT. The basic reason is the wide dynamic of the sputter yield. Not only is the sputter yield a function of primary ion beam parameters as species, acceleration voltage (V_{Acc}) and incident angle; it also depends on sample material, the crystallographic orientation and the composition of the sample (matrix effect).

Composite samples also suffer from selective or preferential sputtering. Low sputter yield elements pile up at the sample surface during the initial sputter phase. The phenomenon decreases depth resolution at interfaces within the sample. Once the sputter equilibrium is reached, this effect can be neglected. Besides the desired sputtering, the primary beam can change the samples lattice structure, chemical composition (implantation) and surface topography, depending on the primary beam characteristics. All of these side effects will have an influence on the measurement result.

In order to increase lateral resolution of the SIMS measurement, the sputter crater area can be decreased. However, this leads to signal degradation as first of all the secondary ion yield decreases with crater area and second of all, the unwanted signal contribution from the crater side walls increases.

By scanning the primary ion beam across the sample, it is possible to generate two dimensional distribution of a specific element at the sample surface. This 2D-approach is limited for classical ion sources as O_2^+ and Cs^+ due to their large ion beam diameter of 100 nm. Although the primary beam diameter can be reduced further by the use of liquid metal ion sources (e.g. gallium), the resulting secondary ion yields are too low for successful element mapping. Some efforts have been done to overcome these limitations by using special analysis test structures [33, 34], excluding applications like FA and reverse engineering.

2.1.3 Electrochemical C-V Profiling

In contrast to the other characterization methods described in this work electrochemical capacitance-voltage profiling (ECV) is rarely used for the analysis of integrated devices. The reason is the extremely large analysis area (and the resulting non applicability of the method for any analysis requiring a spatial resolution below 1 cm) as we will see later on. However, in the PV industry mainstream products like textured silicon solar cells benefit from the features of this analysis method. The heavily roughened surfaces pose serious problems for other dopant profiling techniques. Therefore the ECV technique is a frequently used analysis method in PV industry.

ECV profiling was developed in 1975 by Ambridge and Faktor [35]. Its basic idea relies on a capacitance measurement of an electrolyte-semiconductor-Schottky contact. Depth profiling is achieved through electrochemical etching between the individual capacitance measurements. A detailed review of the method is given by Blood [36].

electrochemical capacitance-voltage profiling obtains the net carrier concentration of the whole sample measurement area. The dynamic range of the technique includes carrier concentrations from 10^{13} to 10^{20} $1/\text{cm}^3$. The achievable depth resolution is 1 nm.

A schematic drawing of the electrochemical cell is shown in figure 2.6. The whole cell is filled with an electrolyte. The sample is pressed against the sealing ring, preventing leakage of the electrolyte. The analysis area is defined by the ring opening. Etching is achieved and controlled by applying a voltage bias between the DUT and the counter electrode. As the dissolution process relies on the presence of holes, these holes have to be generated for n-type material by illumination of the sample. Therefore the cell includes a window.

For the actual $C-V$ measurement a depletion region in the semiconductor

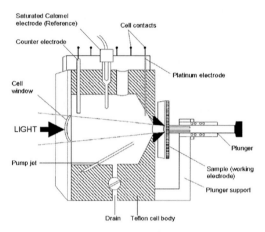

Figure 2.6: *Electrochemical cell for ECV measurements. Etching is done by application of an appropriate potential between the sample and the counter electrode. A depletion region within the semiconductor is generated by a potential between the platinum electrode and the sample.*
(source: PROBion ECVP tutorial (https://www.probion.fr/en/tutorials/ecvp/ecvprofiling.html))

sample is generated by the application of a potential between the platinum electrode and the sample. The applied potential is compared to the potential of a reference electrode. As majority carriers have to be repelled from the sample surface in order to generate the depletion region, for p-type material the electrolyte needs to feature a more positive potential then the DUT. Other way around, an n-type sample needs a more negative electrolyte compared to the sample.

By additional applying of a small AC bias, the carrier density N at the depth of the depletion region (W_d) can by derived from the $C - V$ curves similar to metal-semiconductor systems.

$$N(W_d) = \frac{1}{q\varepsilon_0\varepsilon_r A^2} \cdot \frac{C^3}{\frac{dC}{dV}} \tag{2.9}$$

The correct measurement of the area A is critical for the determination of the absolute carrier concentration.

2.2 2D Dopant Mapping Techniques

As described in the introduction, the request for $2D$ dopant analysis techniques is driven by several developments. One is the continued device shrinking in the high end microprocessor unit (MPU) and memory area. Most of the device features cannot be characterized separately with the aforementioned profiling techniques, just because they are too small. Additionally, the smaller the single device features get, the more import become vertical AND horizontal edges and transitions, only illustratable by $2D$ characterization techniques. The trend towards non planar devices beyond the high end MPUs and memory devices increases the demand for $2D$ dopant analyses.

This section gives an overview of the most commonly used techniques in the semiconductor industry, except for SSRM which is described in detail in chapter 3.

2.2.1 Chemical Staining/Etching

The method of chemically staining or etching a sample in order to obtain information on dopant area formation or even local dopant concentration is known since the early days of semiconductor industry [37, 38, 39]. The method is always used in combination with a microscopy technique as secondary electron microscopy (SEM), transmission electron microscopy (TEM) or atomic force microscopy (AFM) [40] in order to image the staining/etching results. The popularity of the chemical etching/staining technique is due to the availability of both, the required chemistry and at least one of the microscopy methods in basically every semiconductor analysis lab. Both, chemical staining and etching is sensitive to the active carrier concentration of the sample. Therefore the obtained results show the electrical p-n-junction rather than the metallurgical one.

Staining is usually applied to highlight n-doped areas by a selective deposition of a metal (e.g. Cu, Ag, Au or Pt) to these areas. As the staining process covers the whole n-type material, staining is usually used to delineate p-n-junction positions. Although staining works for carrier concentrations as low as $1 \cdot 10^{14}$ $^1/_{cm^3}$ it is difficult to apply at highly doped samples with carrier concentrations exceeding $1 \cdot 10^{17}$ $^1/_{cm^3}$. Moreover, it provides no quantitative results. The spatial resolution is in the range of $20\,nm$.

Dopant area etching is typically done by using a mixture of hydroflouric acid and an oxidizing agent (e.g. nitric acid (HNO_3)). In this way, p-doped areas are preferentially etched. The etching process offers higher sensitivity to differences in the carrier concentration, allowing to approach measurement data also in a basic quantitative way. The method covers carrier concentration ranging from $1 \cdot 10^{17}$ to $1 \cdot 10^{21}$ 1/cm³. In combination with TEM imaging, spatial resolutions below 5 nm could be achieved [41, 42].

Beside the limitations with respect to quantification of the measurement results and the dynamic range, one major issue of the technique is the chemical treatment itself. The results are strongly influenced by the sample surface preparation, the concentration of the applied chemical solutions, the staining/etching time and even ambient lighting (as it has an influence on the carrier distribution near *p-n*-junctions). In conclusion, chemical staining/etching shows poor reproducibility.

2.2.2 Secondary Electron Dopant Contrast

The term secondary electron dopant contrast (SEDC) describes the occurrence of differences in the amount of the detected secondary electrons in an SEM depending on the local doping of the semiconductor sample. The first report of the visualization of dopant areas using an SEM goes back to 1967 [43]. Although even simpler from the point of view of required analysis equipment, SEDC is not commonly used in semiconductor analysis; often the chemical staining or etching described in the previous section is preferred.

The SEDC technique must not be confused with another SEM based method: the SEM voltage contrast (VC). The VC technique relies on the (active or passive) charging of a complete structure [44]. The charging decreases (positive charging) or increases (negative charging) the amount of secondary electron reaching the detector and thus mediate the contrast effect.

In SEDC the contrast mechanism is also obtained by variations in the secondary electron yield at the detector, but these variations are not caused by charging of the structure. Moreover, a charging effect would be counter productive and could screen the desired information.

In fact, the dopant contrast mechanism is not understood completely. Most theories agree in the contrast mechanism being originated from the surface depletion layer, band bending and the presence of surface states at the semiconductor surface [45, 46, 47]. These explanations propose the effect to be of electronic nature, i.e. the measurements correlate to carrier concentration, rather than dopant concentration.

Experimental results show lateral resolution of this method in the range of 10 nm [48]. The dynamic range was reported to be $1 \cdot 10^{16}$ to $1 \cdot 10^{20}$ 1/cm³

[49]. The technique is not limited to test structures, but sample preparation necessary for site specific preparation has a strong influence on the pronunciation of the dopant contrast [50].

The SEM inherent advantages of fast image acquisition and rather simple sample preparation are contrasted by several disadvantages as doping contrast is influenced by experimental factors as magnification, electron dose, V_{Acc}, working distance, extractor voltage and surface preparation.

2.2.3 Scanning Probe Techniques in General

Scanning probe microscopy (SPM) comprise a large number of microscopy techniques. What they all have in common is a physical probe of small spatial extension, that is brought into the near field interaction range of the sample. Depending on the physical properties of the probe, different physical aspects of the sample can be obtained. Once engaged to the sample surface, the probe starts scanning, composing the resulting image line by line. The probe to sample surface distance (or force) is kept constant by the microscope using a feedback loop sensitive to one of the near field parameters.

The first scanning probe microscope was invented by Binnig and Rohrer in 1981 [51]. Their scanning tunneling microscope (STM) already exhibited basically all the elements that are used in SPMs since then. The STM employs a metallic probe tip, commonly made of tungsten, in a very close proximity to a conductive sample surface, in order to achieve tunneling through the remaining vacuum gap [52]. The tunneling current, being strongly dependent on the probe to surface distance is the near field interaction parameter. The microscope drives were realized using piezo actuators, allowing for sub-nm displacement in all three dimensions. Another major aspect in SPM was introduced by Binnig and Rohrer: the feedback loop. In STM the tunneling current (I_t) is constantly monitored and the z-position of the probe is adjusted accordingly to keep I_t constant. The constant tunneling current during scanning ensures a constant probe to sample surface distance, constant tunnel conditions provided and therefore prevents the probe from crashing into the sample. Furthermore the feedback loop allowed to define I_t in order to visualize under different tunnel conditions.

Due to its high lateral resolution (< 0.1 nm) and the outstanding z-resolution (< 0.01 nm), the STM was the first microscope capable of revealing individual atomic surface features in real space [53, 54]. Binnig and Rohrer received the Nobel Prize in physics in 1986 just 4 years after their first publication of the STM [55].

Although the invention of the STM was such a break through, one major disadvantage was its limitation to conductive surfaces. This limitation could be overcome with the introduction of the AFM in 1986 by Binnig, Quate and Gerber [56]. As in STM, AFM uses probes of small spatial extension and the sufficiently fine probe movement is achieved by piezo actuators. The one major difference is the near field parameter that is monitored and kept in feedback loop. Where the STM relies on measuring the tunneling current, in AFM the force between probe and sample surface is obtained. This force is a function of the distance between probe and sample and can be described by the Lennard-Jones potential U_{LJ} (see equation 2.10 and figure 2.7).

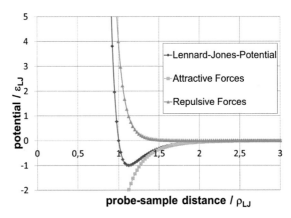

Figure 2.7: *Graph of the Lennard-Jones-Potential (U_{LJ}) including the attractive and repulsive contributions.*

$$U_{LJ} = 4\varepsilon_{LJ} \left[\left(\frac{\rho_{LJ}}{d} \right)^{12} - \left(\frac{\rho_{LJ}}{d} \right)^{6} \right] \qquad (2.10)$$

In equation 2.10, ε_{LJ} is the depth of the potential well of the system and ρ_{LJ} is the distance for a specific system, where attractive and repulsive forces cancel each other out. For long distances d it predicts negative U_{LJ} values, corresponding to attractive forces F. These attractive forces are mainly Van-der-Waals-Forces, proportional to $1/d^6$. For very small distances d the first term proportional to $1/d^{12}$ becomes dominant and results in positive U_{LJ} values equivalent to repulsive forces between probe and sample. The strong repulsion originates in the Pauli repulsion due to overlapping electron orbitals.

For probe to sample force acquisition, the probe is attached to a cantilever and the cantilever deflection is measured. First AFMs used an STM in order to monitor the cantilever displacement. This approach was quickly replaced by optical methods; in the beginning by interference measurements of a laser reflected from the back of the cantilever [57]. Today, the most prominent cantilever deflection measurement method is the beam deflection technique [58]. A focused laser beam is adjusted on the reflective back side of the cantilever. The reflected laser beam is collected using a four quadrant photo diode (4QPD). This is a position sensitive detector. Smallest changes in cantilever deflection can be detected this way. A schematic AFM setup including all major features of the technique as described above is shown in figure 2.8.

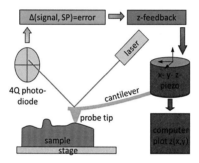

Figure 2.8: *Scheme of the contact AFM setup as used in this work. The cantilever deflection is monitored by the reflected laser beam on a 4QPD. Based on the deviation between detector signal and set point (SP), the z-feedback loop adjusts piezo z-voltage and therefore cantilever bending and probe to sample force. The piezo z-voltage plotted vs x- and y-scanner coordinates results in a topography image.*

Depending on the probe dynamic, two AFM imaging modes can be distinguished: in static mode, as described above, the probe is scanned across the surface while maintaining a constant force, which could be repulsive (contact mode) or attractive (very unstable and therefore hardly used). The second imaging mode is the dynamic mode. Here the probe is actively driven in z-direction at (or near) the resonance frequency of the cantilever. In tapping mode the amplitude modulation caused by the probe sample interaction is used for feedback. The other dynamic mode is the non-contact mode detecting the frequency modulation of the cantilever. As this mode requires very high quality factors for the cantilever oscillation, non-contact mode is usually preserved for AFMs operating under ultra high vacuum (UHV) conditions.

As the STM, AFM proved atomic resolution due to its very high lateral resolution better than 0.1 nm and even better z-resolution of about 0.01 nm [59].

One feature of major importance to SPM techniques is the possibility of probe functionalization and the associated versatility of this class of microscopy techniques. Choosing the right probe enables investigation of the sample in the nm-regime with respect to magnetic (magnetic force microscopy (MFM)), optical (scanning near field optical microscopy (SNOM)), chemical (chemical force microscopy (CFM)), thermal (scanning thermal microscopy (SThM)) or electrical (tunneling AFM (TUNA), conductive AFM (c-AFM) or the later discussed SCM and SSRM) properties.

Especially the electrical characterization modes are of great interest for device characterization. Beside SSRM which is discussed in detail later, SCM should be introduced briefly at this point as it is widely used in semiconductor industry. Other SPM techniques used for dopant visualization and not discussed further here are STM, kelvin probe force microscopy (KPFM), SNOM and scanning microwave microscopy (SMM). Eventually, as mentioned in section 2.2.1, AFM imaging with its superior z-resolution can be used to analyze chemically etched or stained samples.

2.2.4 Scanning Capacitance Microscopy

Scanning capacitance microscopy (SCM) combines an AFM with a high frequency capacitance sensor. An insulating layer (usually native oxide) between the semiconductor sample and the metallic probe leads to a metal oxide semiconductor (MOS) configuration at every scan position. SCM determines the capacity variations of the MOS structure over spatial position.

Interestingly, the first SCM predates the AFM. In 1985 an instrument was build using a capacitance sensor taken from a video disc system and a probe of $2.5 \times 5 \, \mu m$ running in a groove [60]. The first SCM as it is common today using an AFM in combination with a capacitance sensor was patented in 1991 [61].

As the potential of capacitance measurements with high spatial resolution for the characterization of integrated semiconductor devices was obvious from the beginning, efforts have been made to make SCM a quantitative two dimensional carrier profiling technique. Today SCM is one of the favorite 2D dopant analysis methods used in semiconductor industry.

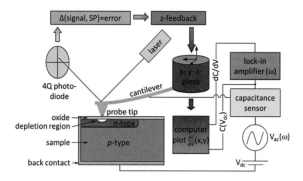

Figure 2.9: *Schematic SCM setup.*

The schematic SCM setup is shown in figure 2.9. The sensor used for capacity measurement has to be very sensitive in order to detect the very small changes in capacity ($\approx 10^{-21}$ F). To enable the electrical measurement in an MOS configuration, conductive probes (most commonly metal probes) have to be used. The probe is electrically connected to the capacitance sensor. The AFM is operated in contact mode, scanning the surface with constant force applied between probe and sample. An advantage of the SCM method are the very small contact forces necessary, as the probe does not need to make a low ohmic, direct contact to the sample in comparison to c-AFM techniques including SSRM. Typical force values are in the nN range. In parallel to the topography image acquisition, at each position the capacitance between probe and sample is measured. An ac voltage of around $\omega \approx 10\,\mathrm{kHz}$ is applied to the probe, inducing a depletion region below the semiconductor surface, modulated with the same frequency. The modulated depletion region results in the measurement of a varying, or differential, probe to sample capacitance. The output signal of the capacitance sensor is a voltage proportional to the measured capacitance. This output voltage is fed into a lock-in amplifier that is referenced to the originally modulated ac voltage. The lock-in amplifier measures magnitude and phase of the sensor output. The magnitude is proportional to the differential capacitance between probe and sample and represents the actual SCM measurement data. Figure 2.10 shows two C-V curves for n-type material. The blue curve corresponds to high, the red one to low carrier concentration. Application of V_{ac} leads to small ΔC and therefore small dC/dV values and vice versa.

From the measurement principle it is obvious, that SCM is sensitive to free carriers rather than dopant atoms. The dynamic ranges from $1 \cdot 10^{15}$

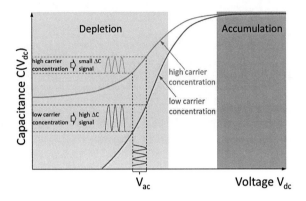

Figure 2.10: *SCM signal generation principle using the example of two C-V curves representing a lowly doped (red) and a highly doped (blue) n-type material.*

to $1 \cdot 10^{20}$ $1/\text{cm}^3$. As all of the SPM techniques, SCM is not limited to test structures. Though, spatial resolution of the technique is limited to approximately 20 nm. This is mainly due to stray capacities originating from the comparatively huge cantilever, the physical separation of the probe and sample and the tip dimension itself. The influence of the cantilever can be reduced by using high aspect ratio probes [62]. Probe and sample are separated by a thin oxide layer. Ideally this is a thin, clean layer exhibiting low density of trapped charge and interface states in order to reduce Fermi level pinning at the surface. The thinner the oxide, the higher the lateral resolution and the signal to noise ratio due to the corresponding local higher capacitance values. Oxides can be grown in different ways affecting quality and reproducibility. The easiest option is to generate a native oxide by exposure of the sample in air for a few minutes. Unfortunately this results in a rather poor quality oxide. Alternative growth procedures for enhanced oxide quality include chemical treatment to oxidizing agents [63], sample exposure by ultra-violet light during baking in air [64] and a baking procedure in an ozone atmosphere [65].

In order to obtain quantitative information on sample dopant concentration, calibration samples of known dopant concentration can be measured and correlated to the DUT data. This demands high requirements on the reproducibility of the sample preparation, due to its enormous impact.

Despite small differences in sample preparation, quantification can also be prohibited by a non-monotonic characteristic of the dC/dV curve. This effect is known as contrast reversal. Sample surface roughness and the corresponding higher density of surface states as well as oxide quality and sample serial resistance could be shown to contribute to this effect [66, 67].

Although its known limitations, SCM is one of the most wide spread 2D dopant analysis techniques in the semiconductor industry.

2.3 Other Dopant Analysis Techniques

This last section is intended to mention some other well known dopant analysis techniques just briefly, as they exhibit serious drawbacks, prohibiting their application to the analysis cases shown in this work.

TEM- or electron holography [68] is sensitive to phase changes caused by electrical or magnetic fields. Therefore, it is able to visualize dopant distributions within semiconductors. The spatial resolution is below 10 nm and the offered dynamic range is 10^{17} to 10^{21} $1/cm^3$. The major drawback of the technique is the required large width of the sample lamella, rendering the electron holography unfeasible for the majority of applications, aimed for in this work.

The atom probe technique [69] determines the original three-dimensional positions of atoms ablated from a sharp tip by application of voltage and/or laser pulses. Atom probe detects dopant atoms rather than free carriers. The method offers superior lateral resolution below 1 nm in all three spatial dimensions. Sensitivity, although, is limited to 10^{18} to 10^{21} $1/cm^3$.

Stripping hall technique [70] combines sheet resistance and sheet hall coefficient measurement for different depths of a sample. Therefore the sample surface is thinned down in between the individual measurement steps. The accuracy and dynamic range of the stripping hall data is outstanding, but the technique is limited to blanket implants, as no spatial features can be resolved.

3 SSRM Methodology

3.1 SSRM as a Secondary AFM Imaging Mode

Scanning spreading resistance microscopy (SSRM) is one of the few analysis techniques offering two dimensional imaging of dopant areas with high spatial resolution and high dynamic range. It was established in the 1990s and can be seen as the combination of the well-established spreading resistance profiling (SRP) method and the relatively new atomic force microscopy (AFM) technique, developed one decade earlier. As it is strongly related to SRP, both share large parts of the theoretical basis.

Although there were similar approaches measuring the current flowing through a probe tip for two-dimensional dopant area characterization [71, 72, 73], it was scanning spreading resistance microscopy (SSRM) that became accepted on a broad basis. The origin of its success was the use of very high contact forces, making the contact resistance negligible and the spreading resistance underneath the probe tip the dominant part in the measurement chain.

SSRM was patented by Vandervorst and Meuris in 1992 [74]. The method generates two dimensional resistance maps of the sample surface, which relate to free carrier concentration and (in equilibrium state) to local dopant concentration in semiconductor material. Implementation of the technique started in 1995 [75] with reports of the first applications in 1998 [76, 77].

A comprehensive review of the analysis method can be found in [78].

SSRM is an electrical measurement technique, as it is sensitive to free charge carriers rather than dopant atoms. Enhancing the SRP method, the AFM approach replaces one probe with a general back side contact, enabling two dimensional visualization of the sample surface with high spatial resolution. As the effective probe radius can be smaller than the actual probe tip radius (see 3.1.2), spatial resolution well below 10 nm can be achieved [79, 80]. SSRM offers a high dynamic range of approximately 10^{14} to 10^{20} $1/cm^3$ in terms of carrier concentration.

SSRM is a secondary imaging mode derived from contact mode AFM.

This configuration has the advantage that topographical information from the z-sensor or z-piezo-voltage can be collected in parallel to the electrical data for a later correlation of both. As a disadvantage of running the method in contact mode, high lateral forces act on the probe tip during scanning, leading to increased tip wear out. As described later, this fact in combination with the high contact forces, necessary for successful SSRM measurements, limits the probe tip choice significantly.

In SSRM a constant dc bias (V_{dc}) is applied between the sample and the conductive probe tip, while the probe is scanned across the sample surface. A logarithmic current amplifier measures the resulting current flowing through the probe tip at every single point of the scanned image. The measurement principle strongly relies on the dominance of the very spreading resistance right below the probe tip at the sample surface compared to the rest of the serial resistance contributions in the electrical measurement path. In this case parasitic resistances can be neglected and the measured total resistance is the same as known from the SRP theory in 2.1.1. In first approximation, the total resistance is equal to the spreading resistance $R(x, y)$

$$R(x, y) = \frac{\rho(x, y)}{4\alpha} \tag{3.1}$$

with the sample local resistivity $\rho(x, y)$ and the tip radius α.

In order to assess the samples local spreading resistance, a sufficiently low contact resistance between probe tip and sample surface has to be established. For that reason, very high contact pressure in the GPa range is required. As a result, contact forces in the µN range between probe tip and sample surface are necessary. These contact forces are several orders of magnitude higher compared to other scanning probe microscopy (SPM) techniques using contact forces in the pN and nN range. As an example, a contact force of $1\,\mu N$ applied to a contact area of $10 \cdot 10\,nm^2$ results in a contact pressure of $10\,GPa$. As shown in detail in subsection 3.1.2 (Nano-Electric Point Contact) there are two reasons for the required high forces. First of all, it is the need for the probe tip to punch through the native silicon-oxide [81], if the sample is measured under ambient conditions or the finally prepared sample was exposed to air. Furthermore, in order to get into sufficiently good electrical contact with the sample, a phase transition of the silicon right beneath the probe tip is required and achieved by the high contact forces [82, 83]. As a consequence of the high contact forces applied for SSRM measurement the probe tips have to exhibit a high hardness. The best probe tips to withstand the arising forces proved to be boron doped diamond probes [84], as discussed in more detail later.

Most prominent applications of the SSRM technique are process characterization [80, 85, 86] and device simulator matching [87]. There is only little literature regarding the application of SSRM for failure analysis [88, 89].

In chapter 4 (Application of SSRM for Technology Development) further applications to silicon solar cells, failure analysis (FA) of silicon integrated circuits (ICs) and characterization of silicon and silicon carbide (SiC) power devices are demonstrated.

As general process characterization and device simulator matching could be performed at suitable test structures, it is the capability to prepare and measure virtually any device of an IC using precise preparation methods like mechanical grinding & polishing or focused ion beam (FIB) that makes SSRM a powerful tool for failure analysis.

3.1.1 Measurement Setup

The SSRM measurement setup is shown in figure 3.1. It is based on the conventional AFM setting. The sample is scanned by a probe with small tip diameter in order to obtain high spatial resolution. The probe itself is attached to a cantilever of a certain spring constant. The x-, y- and z-position and the movement of the cantilever and probe are realized by one or more piezo actuators, providing sub-nm spatial resolution. AFM systems typically use the light pointer principle to control the force between the probe and the sample. A laser beam is focused on the cantilever back side onto a position sensitive sensor. A four quadrant photo diode (4QPD) is used commonly, sensing the deflection of the cantilever in z-direction. The probe to sample force correlates to the sensed cantilever deflection. It also depends on the spring constant of the cantilever, which is discussed later in more detail. In the most common mode of operation (constant force in contrast to constant height mode), the obtained deflection values are compared to a deflection set point value (SP) defined by the operator. The difference between set point and measured deflection is the delta deflection value. A proportional-integral-derivative controller (PID controller) assisted feedback loop adjusts the z-piezo voltage in a way to minimize the delta deflection. By adjusting the probes z-height, the AFM feedback loop keeps the applied force between probe and sample constant, therefore constant force mode. In the classical AFM application, the probes z-height is plotted as function of the x- and y-position of the probe, resulting in a topographical visualization of the sample surface. As the cantilever deflection measurement is very sensitive, resolution better then 1 nm can be routinely achieved by AFM.

The common AFM setup does not involve any electrical measurement of

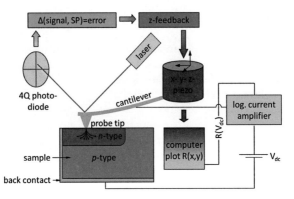

Figure 3.1: *Schematic drawing of the SSRM analysis setup. Piezo actuators, laser assisted beam deflection measurement via four quadrant photo diode (4QPD) and the PID controller controlled feedback loop are common with standard AFM. SSRM specific are the hard conductive probe, the back contact and the logarithmic amplifier.*

the sample. On the contrary, one major stimulation for the development of AFM was to overcome the limitation of scanning tunneling microscope (STM), being applicable to conductive samples only, as it relies on the electrical measurement of the tunnel current between probe and sample. Therefore some specific features of the SSRM setup arise. First of all, the probe tip and cantilever need to be conductive. To ensure current flow, the sample in general and the dopant areas under investigation in specific need a back contact (more details in subsection 3.1.3). Usually the back contact is electrically connected to a conductive sample holder and the holder again to the AFM chuck. The voltage applied between sample chuck and probe tip represents the SSRM measurement voltage V_{dc}.

Finally, an amplifier is necessary for current detection. Whereas the scanner z-position is actively controlled in order to maintain constant contact force, the SSRM amplifier output is a secondary signal. It is passively recorded during scanning, fed into the control computer and displayed as a function of the x- and y-scanner position. As a result, an AFM in SSRM mode allows to obtain topographical information from scanning in contact mode on one channel and the according electrical data on another channel. Beside the requirements regarding the measurement equipment, the sample and especially the sample preparation need to meet some special demands. Enabling the current flow via a back contact is one. Another key for successful SSRM measurement is the preparation of the measurement site. Most often this is a cross section. As will be shown in section 3.2, there are several ways of cross section preparation. All of them target a smooth surface of minimum roughness and with minimum topography features from material differences. Furthermore the measurement surface (prepared or as processed) should not alter the bulk properties of the device under test (DUT) as it does by the presence of an amorphous surface layer or a surface exhibiting a high density of surface states.

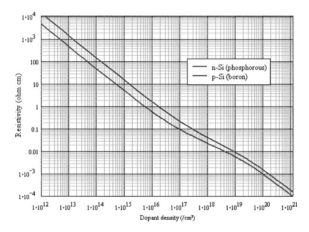

Figure 3.2: *Relationship between silicon dopant density and the resulting resistivity. For the analysis of realistic dopant concentrations ranging from $1 \cdot 10^{14}$ up to $1 \cdot 10^{20}$ $1/\mathrm{cm}^3$, the dynamic of the SSRM method has to cover local resistivity values spanning five orders of magnitude.*
(source: http://www.pveducation.org/pvcdrom/materials/general-properties-of-silicon)

When going more into detail, some more requirements have to be met for successful SSRM analysis. One major requirement affects the probe tip. Besides being electrically conductive, the probe needs to be made of very hard material since very high forces are applied during measurement. The high forces are necessary to punch through the native oxide present at the sample surface when exposed to air and to achieve a low contact resistance between probe and sample (see 3.1.2). More details on probes are discussed in section 3.3, Probes. Another SSRM characteristic relates to the current amplifier. As the local resistivity of the dopant areas to be investigated exhibit a very wide range [15, 16], an amplifier offering a high dynamic range is desirable. Assuming dopant concentration values in the range of $1 \cdot 10^{14}$ up to $1 \cdot 10^{20}$ $1/\text{cm}^3$ present in the DUT, resistivity values span over five orders of magnitude (see figure 3.2).

Although linear current amplifiers can be used (as in conductive AFM (c-AFM) or tunneling AFM (TUNA)), usually logarithmic amplifiers are preferred for SSRM as they offer the required higher dynamic range.

In summary, SSRM is a conductive mode contact AFM technique with some specific characteristics regarding the applied forces, resulting required probe hardness, the dynamic range of the electrical measurement and the sample preparation.

3.1.2 Nano-Electric Point Contact

The goal of SSRM is to determine the very localized and confined spreading resistance right beneath the probe tip. All additional resistance contributions within the measurement path may prevent a correct result. Therefore, before looking at the nano-electric point contact itself in detail, the whole measurement path is examined. The SSRM measurement voltage is applied between the sample chuck and the current amplifier. Serial resistance contributions between those two points include contact resistances from the chuck to the sample holder to the sample back contact and the dopant areas under investigation to the nano-electric point contact itself. In our case, chuck and sample holder are both metallic and tightened by a moderate vacuum. Anyway, if problems regarding a too high contribution in the measurement path exist and the chuck to sample holder transition is under suspicion, both can be glued together by silver paint. One has to be aware, that the application of silver paint induces sample drift during drying of the paint. The same applies for the sample holder to back contact electrical connection. The creation of a low ohmic back contact will be discussed in the next subsection.

Beside the mentioned contact resistances, of course bulk resistances of the different components play a role, too. Metallic chuck and sample holder can be neglected and, if applied correctly, the silver paint too. But there are two contributors, which are capable of seriously influencing the measurement. The first one is the resistance of the silicon between the spot of spreading resistance and the back contact. This current path can get rather high ohmic if a) the silicon material itself has a high resistivity and b) the current path gets very long and c) if the current path is geometrically confined. All three criteria apply for example for the analysis of FinFETs [90, 91].

One serial resistance contribution affecting all SSRM measurements relates

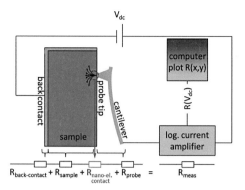

Figure 3.3: *Schematic drawing of all important resistance contributions within the measurement path as described in equation 3.2.*

to the probe tip. Depending on the type of probe, their total resistance values range from about $1\,k\Omega$ up to $100\,k\Omega$. Those values relate to the different types of boron doped diamond tips. Metal probe tips are not considered here, due to their known disadvantages. The influence of probe tip resistance becomes very clear in the Data Calibration section in 3.4.

Summing up the information above, the total measurement resistance can be written as

$$R_{meas} = R_{back-contact} + R_{sample} + R_{nano-el.contact} + R_{probe} \qquad (3.2)$$

In equation 3.2, the typically negligible contributions, like chuck or chuck to sample holder resistance are not considered. Anyway, successful SSRM measurements require that the resistance contribution of the nano-electric point contact ($R_{nano-el.contact}$) dominates the total measurement resistance (R_{meas}). An illustration of the important resistance contribution with in the measurement path is shown in figure 3.3.

Referring to the classical Maxwell ohmic nanocontact model, the resistance of a macroscopic circular contact separating two homogenous conductors can be described as follows (see eq. 3.3) [14]:

$$R_{nano-el.contact} = R_{spread-sample} + R_{spread-probe} = \frac{\rho_1}{4a} + \frac{\rho_2}{4a} \qquad (3.3)$$

Here, ρ_1 and ρ_2 are the resistivities of the materials in contact and a is the radius of the contact area. Obviously, there are two contributions for $R_{nano-el.contact}$. The first one is the sample spreading resistance under investigation ($R_{spread-sample}$). The second one represents the spreading resistance within the probe tip ($R_{spread-probe}$), which can be neglected under most SSRM operating conditions, as $\rho_{probe} \ll \rho_{sample}$.

$$R_{nano-el.contact} = R_{spread-sample} = \frac{\rho_{sample}}{4a} \qquad (3.4)$$

Only in case of very high sample doping ($\rho_{probe} \approx \rho_{sample}$), probe resistance limits the sensitivity to changes in sample resistivity (ρ_{sample}).

Equation 3.4 represents a very straight forward approximation of the measured resistance. Going more into detail, several factors have to be considered to explain observed SSRM measurement results. At first, the Maxwell ohmic contact formula is only applicable if the probe tip radius (a) is large compared to the mean free path of the electrons and holes. For tip radii in the range or smaller than the electrons/holes mean free path, the Sharvin resistance dominates.

$$R_{Sharvin} = \frac{4\rho_{sample}\lambda}{3\pi a^2} \qquad (3.5)$$

As the mean free path depends on dopant concentration, also the transition from Maxwell to Sharvin behavior changes. A more detailed discussion on Maxwell vs. Sharvin resistance can be found in [78]. A comprehensive theory takes both resistance contributions into account.

Aside the differences in the calculated resistance values, both equations exhibit the same basic properties: resistance at the nano-electrical contact increases monotonically with the sample resistivity and decreases with increasing tip radius in contact with the sample.

However, comparison of measured data (IV-curves) with model expectations can reveal significant differences, especially for lowly doped silicon. Reasons can be attributed to influences as pressure between probe tip and sample surface and the presence and density of surface states. Continuing the classical Maxwell ohmic approach those influences can be expressed by a resistance barrier component (see eq. 3.6).

$$R_{spread-sample} = \frac{\rho_{sample}}{4a} + R_{barrier}(\rho_{sample}) \qquad (3.6)$$

Before a more detailed look at the role of the applied pressure and the influence of surface states is taken, one more possible deviation from the theoretical model discussed so far is shown. Up to now only the analysis of homogenous silicon material with a constant resistivity is considered. Practical analysis cases, of course, include spatial variations of local resistivity, due to dopant concentration differences, the presence of highly conductive materials like metals or insulating materials as shallow trench isolation (STI), inter layer dielectric (ILD) or gate oxide (GOX). Those inhomogeneities in the vicinity of the measurement position cause deviations of the measured resistance value compared to the resistance value obtained for homogenous material. A good theoretical explanation including three limiting cases and their illustration is given by De Wolf [92]. He also introduces a correction factor CF accounting for the effects.

$$R_{spread-sample} = CF(a, \rho)\frac{\rho_{sample}}{4a} + R_{barrier}(\rho) \qquad (3.7)$$

One characteristic of SSRM is the use of high probe-sample interaction forces in the μN- or even mN-range, compared to pN- and nN-values typical for other contact AFM modes. Due to the small tip size and the resultant small contact area, local pressure in the GPa range arises. An obvious reason for the necessity of high forces is to punch through the native oxide present at the samples cross section surface, if exposed to air. Eyben et. al. showed that by applying sample surface passivation to prevent the formation of a native oxide, vertical forces can be reduced, but remain in the same order of magnitude [81]. There has to be another explanation for the observed high forces necessary for good electrical contact.

Early in the SSRM method development, a plastic deformation of the silicon material beneath the probe tip was shown [93]. Similar to SRP measurement conditions, the high forces draw responsible for elasto-plastic or even fully-plastic deformation leading to a phase change of the silicon beneath the probe from Si-I phase to Si-II, or β-tin phase [24]. This transition occurs at 12 GPa at purely hydrostatic conditions [94]. If shear stresses are present, for example due to the scanning probe tip in SSRM, smaller pressure (≈ 8 GPa) is sufficient. The β-tin phase is denser with a 22 % volume reduction and exhibits six nearest neighbor atoms (coordination number six) [26]. More significantly with respect to SSRM, the phase transition implies a transition from the sp^3-hybridized diamond structure of Si-I to an sp^3d^2 orbital structure. The electrons of the d^2 orbital are de-localized, leading to a metallic behavior of the six-coordinated β-tin phase. Later studies using molecular dynamic simulations [82, 83, 95] strengthened the hypothesis and revealed the co-presence of two other metastable silicon phases; Si-XIII also with six nearest neighbors and BCT5-Si exhibiting five nearest neighbors. All these high pressure phases show metallic behavior, with the β-tin phase known to be the most conductive. The formation of the Si-II pocket below the probe tip strongly depends on the applied pressure and the probe geometry (as it defines the stress distribution). The mentioned molecular dynamic simulations show the emergence of the metallic pocket not at the sample surface but slightly underneath. With increasing pressure, the pocket growths in size until it reaches the sample surface getting in contact with the probe tip. At this point SSRM visualization conditions are achieved. Further increase of applied pressure leads to dilatation of the metallic β-tin pocket.

The presence of a metastable, metallic, high-pressure phase below the probe tip is a key element of up-to-date nano-electric contact theory [82]. The observed abrupt drop in measurement resistance with increased pressure can be correlated to the metallic Si-II pocket (forming at slightly lower pressure) reaching the sample surface and therefore getting in electrical contact with the probe tip. As a consequence it can be deduced, that probe tip to metallic pocket electrical connection is of metallic character and therefore its resistance contribution in the measurement path is negligible. The dominant spreading resistance appears at the transition of the metastable, metallic region to the bulk silicon material. The β-tin silicon pocket acts as a virtual probe exhibiting significant advantages compared to classical SPM probes. A schematic drawing of the nano-electric contact is shown in figure 3.4. As shown in [82], the size of the pocket can be significantly smaller than the actual probe tip diameter. This is one reason for the superior lateral resolution observed with SSRM compared to other SPM techniques as scanning capacitance microscopy (SCM) or kelvin probe force microscopy (KPFM). Nevertheless, the probe tip radius has an impact on spatial resolution. For larger radii R, the extension of the Si-II phase is increased when reaching the sample surface, leading to a degraded spatial resolution.

The metal-semiconductor interface between the high pressure pocket and the bulk material also implies the nano-electric contact to be of Schottky-type nature [96], in contrast to the hypothesis of an ohmic contact described at the beginning of the subsection. This actually fits much better the observed measurement results. For highly doped material above $10^{19}\,1/\mathrm{cm}^3$ the Schottky barrier becomes negligible and obtained IV-curves show linear ohmic behavior. Not only the rectifying influence of the Schottky-type contact becomes more pronounced for lower doped material, but also the influence of surface states.

The presence of surface states has to be taken into account as the disturbance of the bulk crystal lattice is inevitable at the cross section surface under analysis; no matter whether the sample cross section is prepared in the lab as described in section 3.2 or manufactured in a wafer fab. The

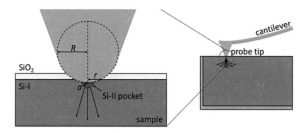

Figure 3.4: *Schematic drawing of the nano-electric contact. The probe tip with radius R penetrates the native oxide layer and indents the sample surface with a contact area of radius r. At sufficiently high pressure a metastable, metallic Si-II pocket of radius a forms underneath the probe tip and acts as a virtual probe. The electrically active radius a can be significantly smaller than the probe tip radius R, enabling the superior spatial resolution of the technique.*

absence of neighboring atoms at the sample surface leads to unsaturated, so called dangling bonds of the surface atoms. Those dangling bonds form two-dimensional energy bands (the surface states) within the energy gap of the semiconductor. An existing surface reconstruction also leads to surface states. Beside these unavoidable intrinsic surface states, there are also extrinsically introduced states, originating from preparation artifacts and residual atoms, most prominently oxygen and carbon. As we will see in the sample preparation chapter, different preparation techniques lead to unequally pronounced influence of surface states. Usually it is aimed for a minimum number of surface states and therefore a minimal influence on measurement data, as surface states lead to a change in carrier concentration at the sample surface or even the formation of an inversion layer [97]. An example for the presence of an inversion layer caused by preparation and the possible access of the non-disturbed sample data is given in subsection 3.2.2 Mechanical Grinding and Polishing. The presence of surface states is immanent to the SSRM method. However, every analysis method applied to cross section surfaces is subject to their influence.

Due to the Schottky contact nature of the nano-electric contact, the SSRM measurement result is strongly affected by the dc sample bias (V_{dc}). Most prominent property is the rectifying or non-rectifying character of the contact, dependent on the polarity of V_{dc}. Although this makes data interpretation more complicated at the first sight, the Schottky like contact behavior can be used as an advantage in some cases to differentiate between *p*- and *n*-doped areas. Therefore the sample has to be scanned with different dc bias polarities.

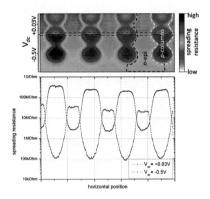

Figure 3.5: *Influence of dc bias a on the example of a super-junction power device. Drastic differences in the obtained spreading resistance are visible, especially for the p-doped columns.*

An example is shown in figure 3.5. Two different dc bias voltages with alternating signs are compared. The image shows a part of the *p*- and *n*-columns of a super-junction (SJ) device. More information on power devices and the application of SSRM to them is given in [98] and chapter 4.3, Power Devices. Main feature of SJ devices is their carrier compensated drift zone, consisting of *p*-doped columns or trenches within an *n*-doped epi matrix. The shown example exhibits a nearly constant *n*-epi doping. The columns are composed of individual, well aligned *p*-doped bubbles. During bottom to top image acquisition, V_{dc} was changed from -0.5 V to $+0.03$ V, resulting in a significant change in spreading resistance values obtained for both, *n*-epi and even more drastic for the *p*-columns. For the negative V_{dc}, the *p*-doped columns show rectifying behavior and the *n*-doped base material is characterized correctly in forward bias direction. For the positive V_{dc}, the resulting SSRM signal shows exactly the opposite behavior. Only the $V_{dc} > 0$ V data reveals the true dopant distribution within the *p*-columns.

SSRM maps local carrier concentration that might be very different from the actual dopant concentration. One possible reason is the application of potential differences within the IC. The measurement of active devices is discussed in detail in chapter 5 (Active Device Characterization by SSRM).

Another reason for non-equilibrium conditions in carrier concentration is the generation of carriers by light. Although often neglected, due to the AFM measurement principle, including a laser beam pointing directly at the top of the cantilever, light induced carrier generation cannot be avoided under standard SSRM conditions. In contrast, the sample can be effectively protected from the ambient light by employing a dark box. Besides that, the ambient light shows a much weaker, if not negligible influence.

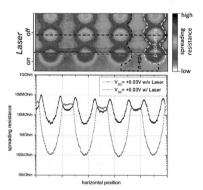

Figure 3.6: *Influence of light induced carrier injection on the example of a super-junction power device. Ambient light was excluded. The AFM laser was the only light source switchable by the dark lift option. In this experiment the spreading resistance maxima at the p-n-junction between the p-doped columns and the n-doped base material is only visible if no carriers are injected by light.*

One AFM used in this work (*Bruker Dimension Icon*) features the dark lift option. In this mode the surface is scanned in contact mode in a first pass, including one trace-retrace cycle. In the second pass the next trace-retrace cycle is scanned at the same z-height trace (an additional z-offset can be applied) as obtained in the first pass. As z-positions during the second pass are determined, the AFM feedback loop and the laser can be switched off; therefore the name, dark lift.

Figure 3.6 shows the second pass of the SSRM image of figure 3.5. In this experiment, V_{dc} is kept constant at +0.3 V and the laser is switched off during image acquisition. Ambient light is excluded. Differences between the areas of constant $V_{dc} = +0.03$ V, but different laser conditions are obvious. Only in dark lift operation, when the AFM laser is switched off during scanning, the expected maxima in spreading resistance, corresponding to the space charge region (SCR) at the p-n-junctions, are visible. Carrier generation by the laser light changes the observed SSRM image significantly. A possible explanation for the observed behavior is the separation of light generated electrons and holes at the p-n-junction between n-epi and p-columns. The holes drift into the p-column area, electrons towards n-epi. This local accumulation results in an increased local carrier concentration and a decrease in measured spreading resistance. The horizontal line scans in figure 3.6 show a decrease in spreading resistance of more than one order of magnitude in the center of the p-column. In parallel an increased carrier concentration can be observed for the n-doped epi material, if negative V_{dc} are used for image acquisition (not shown here). The disappearance of the local maximum in spreading resistance at the p-n-junction can also be attributed to the injection of carriers.

In conclusion it becomes visible, that, although the SSRM approach of measuring the current flow dependent on local carrier concentration and probe tip position for two dimensional dopant area characterization is very straight forward, the actual nano-electrical contact is very complex. It includes the phase transition of the sample material under high pressure and the formation of a virtual probe creating a Schottky-like spreading resistance contact. The size of the virtual probe depends on probe tip geometry and pressure. Furthermore, as SSRM is a surface analysis technique the presence of surface states have to be taken into account. The Schottky like contact makes the correct choice of dc bias important. Also, the often neglected carrier generation due to light has to be kept in mind.

Especially the experiments on signal dependence on V_{dc} and light induced carriers clearly demonstrates the importance of the data evaluation and interpretation. A straight forward evaluation, not taking the discussed, possible occurring effects into account, might result in wrong data interpretation.

3.1.3 Back Contact

The last subsection dealt with the primary contact, intended to define the measurement output. Drawing the parallels to SRP again, the counter electrode, in the case of SSRM the back contact, is also relevant. It requires the electrical conduction of the current flowing through the spreading contact (the nano-electric contact) of all the different dopant areas under investigation to the sample holder. Furthermore, the resistance of the back contact must be significantly lower than the spreading resistance itself. Otherwise the dynamic range would be restricted for material with high dopant concentration by this series resistance contribution. The effect is the same for high-resistance probe tips or other high series resistance contributions in the measurement path. Experience shows the two major sources of elevated series resistance to be the probe tip and the back contact. From measurement data itself the origin cannot be assigned. In this case, a low-resistance reference sample becomes very helpful. Thin film samples of gold and platinum were used in this work. A quick probe tip approach to the metallic sample in SSRM measurement mode shows immediately an elevated probe resistance. Vice versa, if the observed resistance is reasonable low, most probably the back contact resistance is too high. Fortunately, and contrary to SRP and the SSRM nano-electrical contact, there is relatively high freedom of choice for size and location of the back contact. In this work several different types of back contacts were employed in order to meet the different DUT requirements, applied sample preparation and the goal of the analysis.

Every sample needs a global back contact to the sample holder, even in case of advanced analysis using additional local back contacts, described later. Under the assumption of a low-resistive internal current path (through sample material), in the first case, the sample is only front end of line (FEOL) processed (no back end of line (BEOL) metallization). An adequate back contact is usually obtained for those types of samples by simply scratching the sample surface in a safe distance from the cross section to be analyzed using a diamond scriber and subsequently applying silver paint to the fresh scratch. This type of back contact is schematically shown in figure 3.7a. There are other solutions, using an indium gallium eutectic at the opposite side of the analyzed cross section.

The next case features a continuous electrical conducting metallization connected to the dopant areas of interest, as available in most discrete power devices. The same method, using silver paint, with or without the previous scratching, can be applied. Common fully processed, integrated devices exhibit a discontinuous metallization. In this case an electrically conductive layer is deposited over the whole sample area and on top of the clean metal layer connecting and short circuiting the dopant areas of interest. The metallization layer has to be deposited before preparation of the cross section. Possible delamination of the layer in the area of the later sample preparation has to be taken into account. The deposited layer may be connected to the sample holder either by mechanical contact or again by silver paint.

Another approach is realized by employing focused ion beam (FIB) technique. Hereby, a rigid, conductive layer is deposited locally. Usually this approach is used, if all dopant areas under investigation are electrically connected by local BEOL metallization and the preparation process involves mechanical forces, for example as during the grinding and polishing process. Before creation of the local top down back contact, it has to be kept in mind, that significant sample material is removed during the subsequent cross section preparation. Obviously all FIB connected features must be located at the remaining side of the cross section. In order to establish continuous electrical connection to the local back contact, an additional metal film could be deposited globally by sputtering.

In another approach the local FIB deposited metal layer is ensured to cover also a substrate connecting plug or metal line. The substrate then can be globally connected as described in the according paragraph. A schematic drawing of this back contact is shown in figure 3.7b. Advantages of this method are the use of the superior electrical contact to the dopant areas by wafer processing technology, the ability to in situ clean the metal lines or plugs using the focused ion beam before metal deposition and the mechanical stability of the deposited layer. Examples of this type of back contact are given in the sample preparation section 3.2 and in the later application chapter 4. As a disadvantage, this FIB based method is expensive, time consuming and applicable only to small, defined areas.

There are sample configurations prohibiting the creation of a global or local top down back contact. One example are floating dopant areas that are insulated on purpose or as a result of the cross section preparation. A reverse biased p-n-junction causes rectifying behavior, too. However, in some cases existing leakage currents may be sufficiently high to allow a successful SSRM characterization. Even further, and this applies for back contacts in general, if the focus of the analysis is on qualitative, dimensional visualization, rather than quantification, the insulating character of a single dopant area may become of advantage, as it generates sharp transitions in the $2D$ measurement signal.

Current flow through the substrate material can be limited by geometric constrictions, too. The most prominent example is the analysis of FinFETs [90, 91, 99].

Again, it is the application of FIB technique, allowing to electrically conduct local insulated dopant regions. Instead of a horizontal metal bar, a vertical trench or cross section through the area of interest is milled and subsequently covered by the FIB assisted metal deposition. As a drawback, the electrical contact of the FIB deposited platinum to the exposed cross section is of non-ohmic, complex nature, as the cross section surface exhibits surface states and (under normal FIB operating conditions) an amorphous

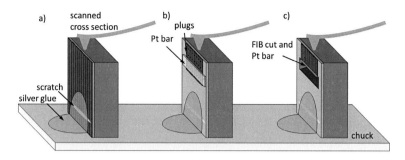

Figure 3.7: *Schematic representation of three different basic types of back contacts. A global back contact is shown in a, a local top down back contact in b) and a local vertical back contact in c). The local (FIB cut and Pt bar) and global features (scratch and silver glue) of the back contacts are not drawn to scale.*

structure. The so prepared local vertical back contact is schematically shown in figure 3.7c.

In summary, there are several methods available for the creation of the necessary back contact. A portfolio covering most of the possible analysis cases, includes simple methods as applying silver paint, diamond scribing followed by the application of silver paint, more advanced methods like global metal deposition by sputtering and more sophisticated techniques as FIB milling and FIB assisted local metal deposition. These methods can be, and in some cases, they must be combined.

3.1.4 Logarithmic Current Amplifier

The experiments shown in this work where conducted using the standard SSRM application module from *Veeco/Bruker*. It includes a logarithmic amplifier offering a measurement range for currents of $10\,pA$ up to $0.1\,mA$. Under test conditions current values as small as $1\,pA$ could be detected. For those small current values the amplifier bandwidth and leakage currents of diodes in the amplifier circuit become increasingly important. The leakage currents being non-logarithmic result in an overestimation of measured resistance. The cut-off frequency of the SSRM application module is specified with $1.4\,^{Hz}/_{pA}$, leading to a slower response to changes of the signal for those small currents. Both effects are shown in figure 3.8. For currents higher $10\,nA$ a bandwidth constantly higher then $14\,kHz$ is available. The impact of bandwidth limitation for small currents is often observed in analyses at material transitions to an insulator, e.g. STI or inter layer dielectric (ILD). Whereas the scan direction from the insulator onto the conductive material shows a sharp transition, the opposite scan direction shows artificial tails as a result of the bandwidth limitation.

On the other side of the scale, minimum measurement resistance is $100\,\Omega$, as there is a serial resistance of this value at the amplifier input present, protecting the amplifier electronics from exceedingly high currents. Under typical SSRM measurement conditions this limitation is not important, as the minimum probe tip resistance available is one order of magnitude higher (approximately $1\,k\Omega$ for CDTP-type probes from *Nanosensors*).

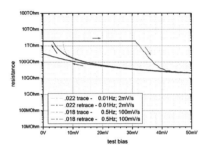

Figure 3.8: *Amplifier output for very low currents (i.e. high resistances) and different sweep rates tested with a $10\,G\Omega$ resistance.*

3.2 Sample Preparation

Sample preparation is key for successful SSRM measurement. This work focuses on the preparation of cross-section samples, as it typically represents the imaging plane of choice. Furthermore, cross section preparation is usually more challenging than the top down preparation approach.

There are some requirements to the SSRM cross section:

- All dopant areas of interest should be included in the prepared cross section and electrically connected to the back contact.

- Topography should be minimal.

- Emergence of contamination layers should be prevented.

- The electrical influence of the prepared cross section on the measurement data should be minimal with respect to the bulk properties.

- From a practical point of view, the prepared cross section should include as much analysis material as possible. As an example, analyses of repeating structures are preferably prepared by cleaving rather than with local preparation methods like FIB.

- From an economic point of view, the cheapest possible method, which is sufficient for target analysis, should be chosen.

The accuracy/inclusion criterion becomes important for small features, perpendicular to the cross section plane, and if additional damage layers, for example due to amorphization, have to be taken into account.

In general, the more pronounced the topographic features, the faster the tip wears out and in worst case the higher the chances of a break of the tip apex. Tip wear out does not just degrade lateral resolution, but also has a direct impact on the SSRM measurement signal. When the contact area increases, the pressure decreases and thus, the sensitive conditions at the nano-electric contact are altered.

The surface roughness of the prepared cross section leads to topographic background noise. For high resolution imaging root mean square (RMS) values below 0.1 nm are targeted and can be achieved demonstrable even by mechanical grinding and polishing. Other extrinsically induced topography features are preparation artifacts. Most prominent example are scratches due to the mechanical grinding and polishing. Further artifacts include debris left on the sample surface, curtaining effect (strongest pronounced at FIB preparation) and cleaving artifacts. The last two mentioned effects are a result of multi material systems and do not occur in homogenous samples.

An alteration of the electrical properties of the surface compared to the bulk electrical data may be caused by surface states or the presence of an amorphous surface layer. In section 3.1.2 Nano-Electric Point Contact the influence of surface states was mentioned. The amount of surface states depends on the preparation technique. As the surface states are present at the sample surface, using higher pressure can overcome their negative effects. The deeper indentation of the probe, plus the increased size in depth of the β-Si phase result in a greater sampling depth. On the other hand, spatial resolution degrades.

Mechanical grinding and polishing and ion beam based preparation generate an amorphized layer at the surface. Dependent on the indentation depth and the size of Si-II pocket this layer contributes to the measurement signal. Examples are given in section 3.2.2 and 3.2.3

3.2.1 Cleaving

Sample cleaving is done by disrupting the sample surface, for example using a diamond scribe, a short distance along the direction of the crystal lattice and subsequently applying stress to both sides of the now damaged crystal in a way the crystal cleaves starting from the damage further in the direction of the crystal plane.

Sample cleaving is the easiest, fastest and cheapest method for cross section preparation. Usually (for limitations see below) the topographic properties are excellent, as the surface is atomically flat. Contamination layers cannot emerge from cleaving. Nevertheless, if done in air, a native oxide will form at the cross section surface and additionally a water contamination layer will coat the surface. Cleaving and further processing under inert gas conditions in a glove box or in vacuum can prevent both. The so prepared surfaces exhibit the best electrical properties compared to the two alternatives discussed later (grinding & polishing and FIB preparation). Nevertheless, a deviation from bulk electrical properties has to be considered, as the surface reconstructs just due to the break of symmetry and the absence of neighboring atoms in one direction. Though, the presence of surface states and the resulting band bending causes far less deviations in the obtained electrical measurement signal compared to the effects induced by mechanical grinding and polishing and ion beam treatment. The generated cleavage can be very long (up to full wafer size), providing plenty of analysis material. Since in practice the cleaving angle is always slightly off the direction of the crystal, cleaving can also be used to prepare smaller, but repeating structures.

The accuracy of the cleaving process is very limited. Purely manual cleaving allows for $< 5\,\text{mm}$ accuracy. Using an indenter or a micro-cleaving tool like the *LatticeAx* from *LatticeGear*[1] enhances accuracy below $50\,\mu\text{m}$. As a consequence cleaving is used for cross section preparation of accordingly large structures with respect to the cross section plane. As the cross section direction is predetermined by the crystal orientation, the flexibility in choosing cross section orientation is very limited using cleaving. Structures rotated with respect to the substrate cannot be prepared by this method, or only with restrictions like a geometrically distorted imaging. Presence

---

[1]LatticeGear, LLC https://latticegear.com/

of front- or back-side-metallization might also prevent using cleaving as method of choice for sample preparation. Contact plugs for example do not cleave in silicon crystal direction and position, but rather end up complete at one or another side of the cross section. This results in strong topographic features. Power technologies commonly use several microns thick front side metallization for current- and heat transport. Those thick metal layers tend to deform during cleaving. Protruding metal can prevent the pyramidal shaped probe tip from acquiring information at the sample edge. An influence of the backside metallization commonly used in discrete power semiconductor devices was observed, as well. Most probably the full wafer backside metallization induces stress in the silicon, causing non ideal cleaving. Another limiting example often found in power technologies (e.g. in recent insulated gate bipolar transistor (IGBT) technology) is the presence of deep trenches. Whether filled with an insulator or poly-Si material, the artifacts generated from the trenches are usually non negligible and other preparation methods have to be applied. In general all features included in the cross section being non crystalline or even worse, exhibiting other crystallographic properties, impede the applicability of the cleaving method.

The topographic artifacts, caused by the back side metallization, can be rather easily bypassed, by chemically or mechanically removing the back side metallization. Although the electrical connection of the substrate now has to be established by diamond scriber scratching and silver paint, the unstructured back side still enables measurement. A simple metal removal is not possible at the sample front side. Within this work a method was developed to locally remove the overlapping metal and preserve the superior electrical properties of the cleaved sample surface. The basic idea is to use the AFM as a microscopic plow. A hard probe is desired, whereas spatial resolution of the tip is not of importance. Therefore an old SSRM probe might be used for this task. In order to protect the cross section surface to be analyzed from material removal, very low deflection values, e.g. probe-sample pressure, are used during scanning. By choosing very high scan speeds above $20\,{}^{\mu m}/_s$ at common gain settings, the feedback loop is not fast enough anymore to follow the abrupt topography changes caused by the metal. As the metal is significantly softer then the probe, the probe withstands the scanning and the metal is removed.

In figure 3.9 the surface modification becomes clearly visible. On the left hand side, the freshly cleaved cross section surface of a discrete power device with a several microns thick metallization is imaged using AFM tapping mode. Strong artifacts caused by the metallization are visible at the top. The silicon top edge cannot be imaged under these conditions. The right

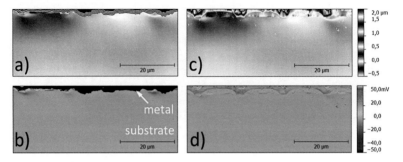

Figure 3.9: *Tapping mode topography a) and corresponding amplitude error b) image of the cross section surface of a discrete power device right after cleaving. Strong artifacts caused by the metallization are visible at the top, preventing analysis of the interesting device region. On the right hand side the topography c) and amplitude error d) images of the same sample after local removal of the metal as described in the text are shown. The silicon edge and the gate structures above become exposed with very little topography difference to the substrate material.*

hand side of figure 3.9 shows the same area after applying the local metal removal technique. Both topography c) and amplitude error d) image show the now exposed silicon edge and the gate structures above. Some small residues can be observed at the microscopically treated surface. As they are lying loosely on the surface, they are only visible during the gentle taping mode image acquisition and removed during the subsequent SSRM measurement.

3.2.2 Mechanical Grinding and Polishing

In contrast to the rather simple cleaving procedure, mechanical grinding and especially the finishing polish step require quite some experience. There are a number of parameters to be chosen correctly, in order to get a sample surface suited for successful SSRM measurement. To list some of the important options, one can use embedded or non-embedded samples, automatic or manual preparation, a huge number of different grinding/polishing pads in combination with just as many suspensions and/or lubricants, all the different contact pressures, different relative speeds between pad and sample and different cleaning procedures.

In this work only non-embedded samples were used. If an additional surface protection was necessary, local platinum deposition by FIB or global covering by a pure epoxy-resin layer or a glass plate attached by epoxy-resin was applied. All samples were manually grinded and polished using rotating disc systems from *Allied*[2] or *Struers*[3].

Although different pads were evaluated, the aluminum oxide and diamond lapping film discs, e.g. from Allied, Struers, Buehler[4], proved to be the best choice. All other pads are significantly softer, and therefore, cause undesired rounding at the transition of materials of different hardness. No additional grinding/polishing suspensions or lubricants other than water were used. Further parameters and handling differ from user to user and rely on individual preferences.

Compared to the FIB technique, mechanical grinding and polishing is still inexpensive. Using grinding and polishing technique, the accuracy for cross section preparation can be significantly improved, compared to cleaving. During this work, contacts as small as 200 nm were prepared in a reproducible and reliable way. Cross sections of a few 10 mm can be prepared at once, enabling the inclusion of multiple analysis sites. The orientation and position of the cross section can be chosen freely, as the mechanical grinding/polishing is not limited by the given crystal structure. In addition, using the hard lapping film discs, it is possible to generate homogeneously flat cross section surfaces, even if different materials are present in the cross section. The presence of front- and backside metallization or deep trenches represents no limitations to the mechanical process.

The accuracy of the method is limited to approximately 100 nm. In case the cross section position needs to be defined more precisely, FIB technique has to applied. Mechanically polished samples normally exhibit artifacts from preparation. Most prominent are scratches in the polishing direction. Despite careful sample cleaning, particles from the mechanical process might be left at the sample surface. Especially the presence of materials softer than silicon (e.g. front- or back side metallization) supports gathering of residual particles. The high probe to sample contact forces used in SSRM have a positive impact at this point, as most of the particles are easily pushed aside and existing scratches can even be leveled out to some extend during scanning. However, the higher the contact forces to be used, the more the spatial resolution is degrading.

---

[2] *Allied M − Prep* 1 - Allied High Tech Products, Inc. - http://www.alliedhightech.com
[3] *Struers TegraPol* 15 - Struers - http://www.struers.com
[4] Buehler http://www.buehler.co.uk

Using this type of preparation, surface contaminants cannot be avoided. The surface is in contact with air, water or other lubricants and the lapping film disc consisting of the aluminum oxide- or diamond particles embedded in a resin matrix. The coarse interaction at the sample surface in this mechanical process also leads to sub surface cracks, the incorporation of oxygen and amorphization of the sample surface [100]. Except for the superior planarization of different materials within the cross section, the general topographical and electrical properties of mechanically polished surfaces are significantly worse than for cleaved samples.

If surface related effects, as the presence of an amorphous layer, incorporation of impurities and a general increase of surface states negatively effects the measurement signal, hypothesis is, a deeper sampling of the virtual probe by applying higher probe to sample contact forces (F_C) might probe through the electrical artifact layer. Figure 3.10 shows two SSRM depth profiles through the p-column of a SJ device obtained at $V_{dc} = +0.5$ V. The initial spreading resistance curve is obtained at medium contact force ($F_C \approx 59 \,\mu$N). A second curve is obtained at very high contact forces ($F_C \approx 306 \,\mu$N) and an additional disabling of the slow scan axis for 20 lines. Disabling of the slow scan axis leads to a repeated line scan at the same position, in our case associated with additional sample material removal. The contact forces were deduced from scanner height values and the nominal spring constant of the probes used[5], even though the observed change in scanner height includes cantilever bending and probe intrusion into the sample surface. When slow scan axis is disabled, but contact force is kept constant, the difference in scanner height is only a result of sample material removal.

Beside the lower overall resistance for the high contact force measurement (due to the increase of the virtual probe), a contrast reversal back to the trend fitting expectations and known from the cleaved sample can be observed. Maxima in dopant concentration in the center of the bubbles now correspond to minima in spreading resistance. Additionally the p-n-junction between the lower end of the p-column and the n-epi becomes visible as a local maximum in spreading resistance. Although the influence of the inversion layer, resulting from mechanical polishing can be eluded, using very high contact forces brings serious draw backs, like fast probe tip wear out, reduced lateral resolution and sample material removal. In case of disabling the slow scan direction during data acquisition, two-dimensional imaging capability is lost completely.

---

[5]$NCHR$-type probes from *Nanosensors* (nominal spring constant 80 N/m)

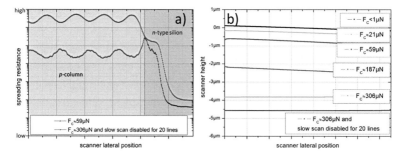

Figure 3.10: *SSRM line scans along the center of an individual p-columns of an SJ MOSFET a) imaged at moderate contact force $F_C \approx 59\,\mu N$ and very high contact forces $F_C \approx 308\,\mu N$ and additionally disabled slow scan direction. Scanner height values for the two cases and additional contact forces are shown in b).*

3.2.3 Focused Ion Beam

Beside secondary electron microscopy (SEM), focused ion beam (FIB) technique is one of the working horses in physical failure analysis (PFA). The use of ions instead of electrons for the primary beam enables or enhances different contrast mechanisms in particle beam imaging. FIB for example shows very strong channeling contrast in crystalline or polycrystalline material and enhances voltage contrast significantly. Due to the high mass of the ions in comparison to electrons, FIB imaging exhibits a significant sputtering effect of the sample surface, with sputter rate being dependent on primary beam ion species, sample material and composition and acceleration voltage (V_{Acc}). Sample sputtering might be a disadvantage for imaging purposes, but enables sample preparation and modification. Advanced tools feature an electron- and an ion gun, in order to observe FIB cut cross sections in situ by simultaneous or successive electron beam imaging. Equipped with the right gas injection system (GIS) it is possible to not only remove material with high accuracy, but also deposit material, for example a local protection layer or microscopic electrical connections.

The major advantage of FIB is its unsurpassed accuracy. Depending on beam aperture size, the FIB cross section position can be determined well below 100 nm. Besides its high accuracy, FIB offers excellent control over processing parameters, and therefore, increases repeatability. Although single failing memory cells, e.g. static random access memory (SRAM) or read only memory (ROM) as presented in chapter 4, could be cross sectioned under ideal conditions using mechanical grinding and polishing, the risk to irrecoverable damage the target device by a bigger scratch or chipping is not negligible. Whenever specific structures smaller 500 nm have to be cross sectioned, FIB is the method of choice.

In comparison to both mechanical methods discussed above, the FIB equipment is very expensive. Furthermore only microscopic cross sections of several 10 μm width and depth can be prepared effectively. Larger areas would require an excessive amount of processing time, associated with increasing sample drift or the necessity to use higher beam apertures, both limiting accuracy.

The major drawback of all ion beam based methods is the amorphization of the cross section surface and the parallel incorporation of primary beam particles. Both, thickness of the amorphous layer and depth of particle implantation strongly correlate with the applied acceleration voltage V_{Acc}. Transmission electron microscopy (TEM) and time of flight secondary ion mass spectroscopy (ToF-SIMS) investigations of an FIB prepared cross section surface are shown in figure 3.11 [101]. For the common combination of gallium ions and $V_{Acc} = 30$ keV the resulting silicon amorphous layer is approximately 20 nm thick. Gallium is present even 10 nm below the amorphous layer. Most investigations of FIB induced artifacts were done with regard to TEM lamella preparation. There are only a few studies on the influence of FIB preparation on SCM and SSRM measurement [102, 103, 104]. Furthermore, the cited articles investigate only top down FIB modifications, but not cross section preparation. Nevertheless, the presence of an amorphous surface layer, implantation of primary beam particles and the associated creation of surface states leads to an altered sample surface, and therefore influences SSRM measurement in a detrimental way.

Using lower acceleration voltages for final cross section preparation reduces the damage effects. However, the performance of common ion columns strongly degrades for V_{Acc} smaller 5 keV. Alternatively, an $ex-situ$ finishing step using a broad ion beam tool can be introduced (e.g. $Leica EM - RES101$). Some broad ion milling tools enable acceleration voltages down to 20 eV, reducing the amorphous layer drastically [105]. Furthermore, the broad ion milling tools usually employ argon instead of the metallic gallium.

Figure 3.12 points out the damage effect due to ion beam preparation. The imaged sample is an n-type calibration sample, consisting of several planar layers of constant doping concentration. The dopant concentration of the respective layers increases stepwise towards the sample surface. The

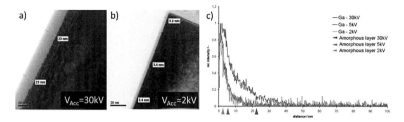

Figure 3.11: *The amorphous layer at the surface of a cross section prepared by FIB using standard acceleration voltage (V_{Acc}) of 30 keV is shown in a). In comparison, b) shows a significantly reduced amortization layer thickness, due to the application of a reduced V_{Acc} of 2 keV during FIB cross sectioning. The depth of Ga implantation depending on acceleration voltage was measured by ToF-SIMS and is shown in c).*

range of dopant concentration spans approximately four orders of magnitude. Actually, the sample is not prepared by focused ion beam using gallium ions at V_{Acc}= 30 keV, but broad ion beam milling with argon ions and only V_{Acc}= 6 keV. Two successive scans were carried out at the same location. The first SSRM image (a) shows weak contrast, confirmed by only two orders of magnitude range in the spreading resistance signal of the silicon (red colored line scan at the right hand side). The second image (b), shows an enhanced contrast at the same color coding. Accordingly, the range of the spreading resistance signal is now strongly enhanced to the expected approximate four orders of magnitude. Interestingly, the highly doped layers at the top of the sample show a decreased conductivity and the lowly doped substrate an increased conductivity during the first scan compared to the second scan. Both can be explained by the presence of an amorphous surface layer. The improvement in measurement range can be attributed to material removal of the topmost damage layer by FIB during the first scan.

Although several attempts were done using low energy FIB and broad ion beam to clean the final cross section, artifacts like reduced signal range and screening of lowly p-doped areas were still present. This indicates that surface states are still remaining, although avoiding an amorphous surface layer or removing it.

Additional artifacts during FIB preparation might be introduced for non-homogenous samples exhibiting local differences in sputter rates, e.g. presence of contact plugs. This leads to a pronunciation of topography at material transitions, especially in parallel direction to the ion beam. The

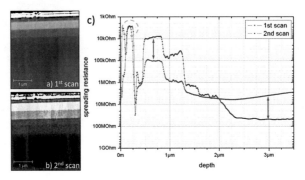

Figure 3.12: *Comparison of first a) and second b) SSRM scan of an ion polished sample. Profiles obtained from both images are shown in c). An increase in contrast of the SSRM signal is obtained from the first to the second scan. This improvement can be explained by removal of topmost damage layer material during the first scan. For comparison, a metal layer deposited on top of the sample exhibits unchanged low spreading resistance values (green circle).*

effect is known as curtaining [106] and leads to accelerated tip wear out and topography induced artifacts in the SSRM image.

Within this work, a preparation procedure was developed, preventing FEOL induced curtaining artifacts. A 130 nm complementary metal oxide semiconductor (CMOS) technology node sample was used for the study. Common FIB cross sectioning parameters were used: gallium ions, $V_{Acc} = 30 \, \text{keV}$ and constant beam current. The topography artifacts due to curtaining were evaluated for different incident angles of the ion beam. Figure 3.13 summarizes the results. At first a standard cross section was prepared (figure 3.13 a), thereby the ion beam hits the contact plugs (low sputter rate) and the ILD (high sputter rate) before the silicon. The difference in sputter rates in combination with the sharp edges result in a strong curtaining effect with steep topographic steps within the silicon material. In a second experiment, the ion beam was tilted approximately 15° with respect to the sample surface normal (figure 3.13 b)). The SEM image shows significantly reduced curtaining effect. The AFM data confirms the topography differences to be smaller and the slopes of the steps to be shallower, but both still being significant. In a third experiment, the incident angle was drastically changed by 180°, to cut from the silicon bulk material towards the FEOL metallization. Figure 3.13 c) shows the result. Artifacts from curtaining

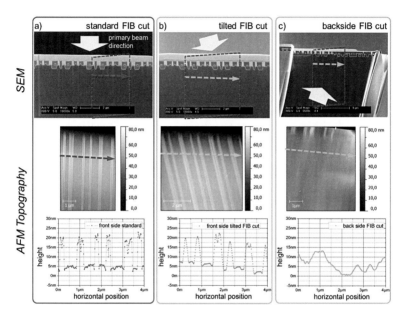

Figure 3.13: *FIB cross section preparation of a* 130 nm *CMOS device using different incident angles for the primary ion beam. At the top, three SEM images of the standard FIB cut a), the tilted top down FIB cut b) and the backside FIB cut c) are shown. AFM topography images of the three different FIB preparation methods are shown below. In the last row the corresponding horizontal topography profiles are shown. The backside FIB cut shown in c) completely avoids curtaining in the bulk silicon material.*

are visible in the upper metallization and ILD layers, but completely absent in the silicon material. As a result, SSRM imaging artifacts due to FIB curtaining can be completely avoided. Furthermore, tip wear out can be significantly reduced.

3.3 Probes

Due to the measurement principle and the physics of the nano-electric point contact, requirements for the applied probes are: high conductivity, high hardness and small tip radii. Very early in the development of SSRM, boron doped diamond probes showed to deliver the best compromise [93, 75, 72]. Even though solid metal probes offer superior conductivity, the metal tip does not withstand the high contact forces and wears out quickly [107, 108]. As a result the applicability of metallic probes is very limited. One example using tungsten probes for FA is shown in [109]. On the other hand the resistivity of diamond can be lowered by boron doping only to a certain degree. Commercially available probe tips exhibit resistivity values two to five orders of magnitude higher than those of metallic probes.

3.3.1 Diamond coated Probes

Two basic types of boron doped diamond tips can be distinguished. The first ones use a standard silicon probe as basis and apply an additional boron doped diamond coating by sputtering (for example *Nanosensors CDT-* or *CDTP*-type probes[6]. SEM images of *CDTP*-type probe are shown in figure 3.14. In a) the underlying geometry of the silicon cantilever and attached probe tip are shown. Even at this low magnification, individual diamond grains are visible. The grain structure becomes clearly visible in figure 3.14 b). This image illustrates two of the biggest disadvantages of the sputtered diamond approach. The overall tip radius increases (specified for example to 100 to 200 nm for *CDT*-type probes). Under ideal conditions, only the most protruding grain acts as a nano tip. Nevertheless, the arrangement of the grains is random, leading to a possible multi tip imaging. The risk of double- or multi tip artifacts within the SSRM image increases with increasing sample topography and applied contact force.

Tip wear of diamond coated probes usually happens by the rupture of single diamond grains, leading to sudden offsets in the SSRM image. As

---

[6]*CDT* stands for *Conductive Diamond coated Tips*. The *P* in *CDTP* reflects an extra (*Plus*) coating for higher conductivity.

the pressure between tip and sample depends on the contact area, changes in grain arrangement at the tip usually lead to a change in the observed spreading resistance signal, too. Figure 3.14 c) and d) show a worn off probe tip. In c) the silicon pyramid underneath the diamond coating is visible. In case the silicon is the most protruding part of the probe, contact resistance will increase significantly. Figure 3.14 d) is obtained at an angle of 20°, showing a plateau with significantly increased contact area at the tip position.

In exchange for the structural disadvantage of the diamond coated tips, those probes offer the lowest resistance values with respect to the diamond based probes that are used for SSRM. The extra thick coated $CDTP$-type probes from $Nanosensors$ exhibit resistance values down to $1\,\mathrm{k\Omega}$.

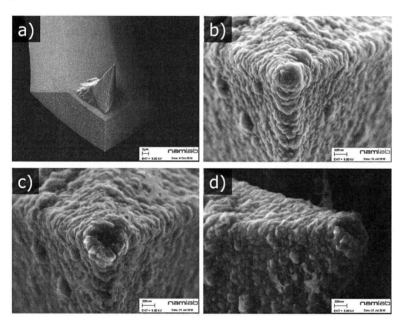

Figure 3.14: *SEM images of a diamond coated probe tip. Diamond is sputtered onto a standard silicon probe a), leading to randomly arranged grains b). The same tip after SSRM measurement is shown in c) and d). In c) the silicon material underneath the diamond coating is visible. Image d) illustrates the increase in contact area.*

3.3.2 Pyramidal moulded Diamond Probes

The second prominent probe type used for SSRM are pyramidal moulded boron doped diamond probes, developed at *IMEC* [84, 110, 108]. The diamond material is deposited into an anisotropically etched silicon form. After cantilever formation, the silicon mold is etched and the diamond pyramid remains. The biggest advantage of these probes is their well-defined geometry, preventing double- or multi-tip artifacts. Furthermore, compared to the diamond coated tips, the lower aspect ratio and the full diamond structure (no silicon core) increase life time. As a disadvantage, the resistivity of the probes is one to two orders of magnitude higher (10 to 100 kΩ overall resistance) than that of diamond coated tips. Figure 3.15 shows an optical microscopy image a) and an SEM image b) of the diamond tip attached to the metal cantilever. In c) the well-defined pyramidal structure of the moulded diamond is visible.

Both types of probes offer different advantages and complementary disadvantages. Application cases with relaxed spatial resolution requirements (≥ 20 nm) but the necessity to electrically resolve differences in or between highly doped areas, can be addressed by the generally higher conductive diamond coated Tips. If best spatial resolution is required, the moulded diamond tips show their predominance.

3.4 Data Calibration

As discussed earlier, SSRM measurement data represents the local spreading resistance of the sample, but not the actual dopant concentration. In many cases this information is sufficient for analysis. In failure analysis for example, often comparative measurements of the failing device versus a reference or pass device are performed. Existing differences in implant area dimensions (see section 4.2.1) or blocked implants (see section 4.2.2), to name only two, might fully explain the electrical failure of the IC.

On the other hand, process characterization and device simulator matching usually require the measurement data to be calibrated to actual dopant concentration values. As in SRP, calibration is done by comparison of the measurement data to the data obtained from a known sample. As a consequence, additional measurements are necessary. Measurement parameters as contact force and scan speed, but most important, probe tip condition need to be constant at both, measurement of the DUT and the reference.

Very early in SSRM method development, special calibration samples were

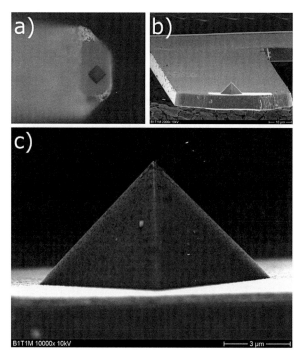

Figure 3.15: *Pyramidal moulded boron doped diamond tip attached to metal cantilever, a) optical microscopy, b) SEM. An SEM image obtained at higher magnification shows the well-defined pyramidal structure of the moulded diamond c).*

used [18, 111]. All of them feature several distinct planar layers of different doping concentration. Thus, the number of additional measurements for data calibration is reduced to a minimum. In this work, calibration samples from *IMEC* were used, whereby *n*- and *p*-dopant calibration was done using different samples. Each sample was part of a wafer, characterized by secondary ion mass spectroscopy (SIMS) and SRP. The dopant concentration profile of the *n*-calibration sample as obtained by SIMS is shown in figure 3.16.

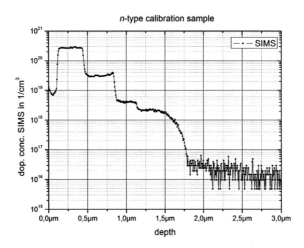

Figure 3.16: *SIMS profile of the n-type calibration sample. The individual layers span a dopant concentration range from approximately* $2 \cdot 10^{16} \, 1/\mathrm{cm}^3 \, to \, 3 \cdot 10^{20} \, 1/\mathrm{cm}^3.$

Figure 3.17 shows the basic steps of the calibration flow using the example of the SSRM analysis of an laser doped selective emitter (LDSE) processed solar cell. For more information on the specific application case, see 4.1.1. In the first step, the DUT is scanned (upper left image). In the following step, the same measurement parameters including the same probe tip are applied to obtain corresponding SSRM data of the calibration sample. In this case only the *n*-type sample is applied, as the area of interest is the *n*-type emitter of the solar cell. Both $2D$ images feature the same color coding. In this experiment, amplifier output values were used, rather than corresponding resistance values. Both are correlated by $R = 10^{V_{AmpOut}+6}$ for negative sample bias values (V_{dc}). In the third step, the one-dimensional profile of the SSRM data is fitted to the known dopant profile of the calibration sample. By fitting SSRM profile data to the SIMS profile, one assumes a 100% activation of dopant atoms, which might lead to errors, especially for the highly doped regions. The fitting process is described in more detail below. As a result of fitting both, the calibration sample SSRM data to the corresponding SIMS data, a transfer function is received (step 4). This transfer function is applied to the initial DUT SSRM data. As result, a $2D$ dopant concentration map (for n-type material) of the LDSE is obtained (5).

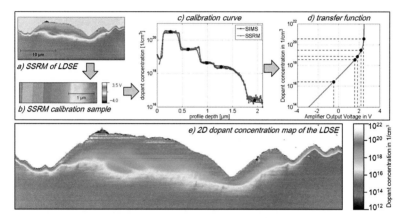

Figure 3.17: *Data calibration flow using the example of a solar cell LDSE. a) Imaging of the DUT. b) Imaging of the calibration sample. c) SSRM data fitting to the known SIMS data. d) Generation of the transfer function. e) Application of the transfer function to the initial DUT data to obtain $2D$ dopant concentration map of the device.*

The basic idea of the fitting procedure used in this work is that measured resistance value consists of two parts: the actual spreading resistance of the sample material and an additional serial resistance contribution from the probe tip. Figure 3.18 shows the effect of the different fit parameters. The abscissa represents amplifier output voltage. Applying the transfer function of the logarithmic amplifier ($R = 10^{V_{AmpOut}+6}$) leads to the black curve, linking V_{AmpOut} to the measured resistance R. The right ordinate displays SIMS measured dopant concentration linked to the left ordinate by $N_D = 10^{23} - R_{spread-sample}$. A calibration factor cf is introduced shifting the curve to higher sample spreading resistance values for constant V_{AmpOut}

(blue curve). While the factor depends on the rather arbitrarily chosen 10^{23}-term introduced above, changes in this factor might be attributed to variations at the nano-electrical contact, for example the contact area a (see equation 3.3). Applying this simple shift, it becomes visible, that the first three data points from the calibration sample ($2 \cdot 10^{16}$ $^1/\mathrm{cm^3}$ to $4 \cdot 10^{18}$ $^1/\mathrm{cm^3}$) can be fitted by the logarithmic amplifier transfer function. $1\,\mathrm{V}$ increase in V_{AmpOut} correspond to one order of magnitude lower measurement resistance and one order of magnitude higher dopant concentration.

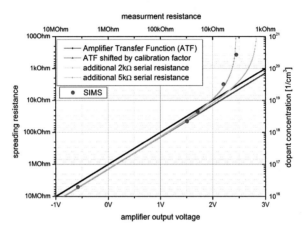

Figure 3.18: *To fit the data points obtained at the calibration sample to the SIMS values, the amplifier transfer function (black) is multiplied by a calibration factor (blue). Two additional serial resistances ($2\,\mathrm{k\Omega}$ (cyan) and $5\,\mathrm{k\Omega}$ (green)), reflecting the probe tip resistance, were added. The $5\,\mathrm{k\Omega}$ curve fits well the SIMS characterized calibration sample data.*

Higher dopant concentration values show a steeper slope of the corresponding amplifier output values. Here, the influence of the probe tip serial resistance is not negligible anymore. Two curves with different serial resistances are shown to illustrate the effect. Assuming a $5\,\mathrm{k\Omega}$ serial resistance R_{serial}, the data points obtained at the calibration sample can be well fit. The resulting transfer function to be applied to the n-type material of the initial DUT SSRM data can be written as:

$$|N_D| = 10^{23} - ((cf \cdot 10^{(|V_{AmpOut}|+6)}) + |R_{serial}|) \qquad (3.8)$$

While the whole calibration procedure can be done manually, within this work a *Matlab* program was developed, to fit the SSRM calibration curve to the SIMS data. The algorithm automatically adjusts data offset and possible distortion in depth (e.g. due to drift). Core feature is the determination of the calibration factor cf and serial resistance R_{serial} to automatically generate the transfer function and to apply it to the initial DUT SSRM data.

4 Application of SSRM for Technology Development

This chapter presents a selection of the analyses by means of scanning spreading resistance microscopy (SSRM) conducted during this work. The applications are divided by technology. The first section discusses the application to photovoltaic (PV) devices, specifically the process characterization of front side selective emitter structures. In the second section, localized deviations in dopant concentration in integrated circuit (IC) complementary metal oxide semiconductor (CMOS) technology are discussed for two examples. The third section introduces the application of SSRM to discrete power devices. In this section the applicability of the analysis method to another material system, silicon carbide (SiC), is demonstrated, too.

4.1 Photovoltaic

Since the development of commercially available solar cells in 1954 [112], the technology was used for the energy supply of autarkic systems, e.g. satellites. Since the last two decades the whole photovoltaic market has grown exponentially. The development is driven by the strive to end the dependency from nuclear and fossil-fuel energy. Within the growing world market, silicon based solar cells show a market share of 90 %[1].

In silicon based solar cells, current is generated, as light injected carriers are separated by the built-in electric field of a p-n-junction. This junction between base material and the so called emitter is of great interest for technology development and process characterization. Beside one dimensional profiling methods like electrochemical capacitance-voltage profiling (ECV), SSRM was successfully applied for the analysis of silicon solar cells front side emitter [113]. Furthermore, SSRM shows the potential to investigate specific technology features like back surface field (BSF) [114, 115] or back contacts [116].

---

[1]Photovoltaics Report 2014/2015, from *www.ise.fraunhofer.de*

For silicon solar cells the emitter process is very critical. In standard solar cells there is one emitter type all over the cell. The requirement for good spectral response and a low resistance connection to the front side metallization leads to a trade-off situation. An emitter showing good spectral response leads to high resistance values. This would negatively influence the electrical connection to the front side grid and therefore lead to decreased cell efficiency. A very low ohmic emitter on the other hand connects very well to the front side metallization, but shows weak spectral response, also resulting in non-ideal efficiency of the cell. The engineering task using one emitter type across the whole solar cell is to find the emitter with the optimum combination of spectral response and electrical contact to the front side grid and thus allowing best cell efficiency. One concept to further increase the efficiency of solar cells is to introduce the selective emitter concept [117]. Thereby the emitter is split into two parts. One part is located between the front side metallization. As this part is exposed to the incident light, it exhibits good spectral response. The poor resistance values are rather irrelevant, since this emitter type does not contact the front side metallization. The second emitter type is located directly beneath the metal grid and therefore needs to be low ohmic. Although this leads to a worse spectral response it does not matter because this emitter type is not exposed to the incident light. Selective emitter characterization is not straightforward, because standard analysis techniques like spreading resistance profiling (SRP), ECV or secondary ion mass spectroscopy (SIMS) do not offer the needed lateral resolution or struggle with the textured solar cell front side [118]. SSRM is one of the rare analysis tools enabling the required characterization.

4.1.1 Laser Doped Selective Emitter

The samples for this analysis were kindly provided by Roth & Rau AG[2]. The solar cells were made of mono-crystalline silicon material. In the laser doped selective emitter (LDSE) approach a shallow lightly doped emitter (regular emitter) is created at the whole solar cell surface in the first step. This emitter has an enhanced spectral response compared to the emitter from a standard process. In a second step an additional, heavily doped emitter exhibiting a low sheet resistance is added by the laser doped selective emitter process. Due to the following self-aligned light induced plating process, the selective emitter is located exactly at the position, where the low sheet

---

[2]http://www.roth-rau.de/

resistance is needed; below the front side metallization. That leaves the areas between the metal fingers where the light falls in with the regular emitter exhibiting the good spectral response. Splitting the emitter doping in a regular and a selective emitter as described, leads to an enhanced spectral response of the cell and a reduction of the metal grid finger width. In summary, this leads to an increased cell efficiency.

The whole process flow is shown in figure 4.1. It is possible to implement the LDSE technology in an existing standard process flow, by adding only two additional process steps: the laser doping itself and the following plating of the front contacts.

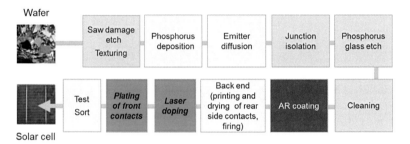

Figure 4.1: *LDSE process flow. Only two additional process steps (green blocks) are necessary compared to an existing standard process flow: the laser doping itself and the following plating of the front contacts.*

Figure 4.2 shows the LDSE process itself [119]. In a) from bottom to top there is the back side metallization, the (green colored) back surface field (BSF), the p-doped bulk silicon and at the front of the cell (red) the regular emitter from the first emitter doping step and the passivation layer (blue). The laser doping process starts in b) with the deposition of a thin phosphorous rich layer. Then follows the actual laser diffusion c). A laser is scanned across the cell surface, locally removing the passivation layer and melting the top region of the bulk silicon. During this step phosphor from the previously deposited dopant layer is introduced in the molten silicon. A cleaning step follows d), removing the rest of the phosphorus layer and preparing the grooves for the light induced plating of the front side metallization e). As a result of the high conductivity of the LDSE below the metal fingers, serial resistance is reduced, enabling thinner metal fingers and therefore a reduced shading.

Due to the textured front surface and the presence of metal lines, the

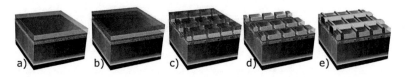

Figure 4.2: *LDSE doping process. The single process steps are described within the text.*

sample cross section was prepared by mechanical grinding and polishing. Prior the actual cross sectioning, the sample front side was protected by sputtering a 500 nm thick SiO_2 layer on top of the surface. Additionally, a glass plate of 1 mm thickness is glued on top using an epoxy resin[3] exhibiting a low viscosity, so even small voids at the sample surface are filled. Both, the SiO_2-layer in combination with the epoxy resin and the covering glass plate, improve the stability of the otherwise very brittle pyramids of the textured solar cell surface significantly. After curing of the epoxy-resin for about 12 hours, coarse grinding started with silicon carbide discs of grit 500 up to 1400. For polishing, diamond foils were employed, reducing the diamond grain size down to 100 nm for the final polishing step. The preparation progress was monitored by optical microscopy and stopped as soon as no further improvement of the sample surface could be observed. Silver glue was used to electrically connect the metal grid fingers and the sample back side to the metallic sample holder.

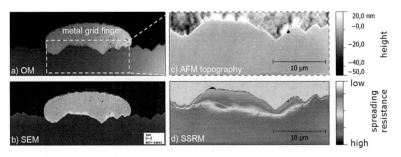

Figure 4.3: *Cross section of an LDSE prepared by mechanical grinding and polishing.*

SSRM was applied for the characterization of the LDSE with respect to

---

[3] $Epo-Thin$ from *Buehler*

its lateral and vertical dimensions and its dopant concentration distribution. Commercially available boron doped polycrystalline sputtered diamond tips (*Nanosensors CDTP-NCHR*) were used for imaging. All measurements were done using a sample bias $V_{dc} = -50\,\mathrm{mV}$, whereas the tip was kept at virtual ground. One of the metal lines was chosen and imaged (figure 4.3) using optical microscopy (OM) a) and secondary electron microscopy (SEM) b) first. The topography channel of the successive SSRM measurement is shown in c), the SSRM data itself in d).

As shown in figure 4.4, a $2D$ dopant concentration map of the LDSE is obtained after data calibration (3.4). Clearly visible are the substrate, the regular emitter and the selective emitter in the center of the image. The passivation layer is indicated as thin dark line right below the metal grid finger (white). Having a closer look, one can see breakthroughs in the passivation layer (marked by ellipses), showing the actual contact spots of the metallization finger to the emitter. The area of the selective emitter is obviously visible. Due to the high spatial resolution of the technique, it is possible to extract accurate information regarding the width and depth of the selective emitter.

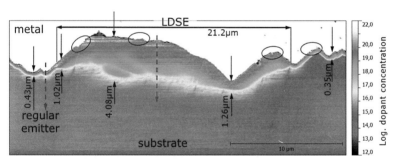

Figure 4.4: *SSRM image of the LDSE including dimensional measurements. Depth and width of the selective emitter can be extracted precisely. The blue and red dashed arrows mark the positions of the extracted depth profiles shown in figure 4.5.*

Even further, depth profiles of the regular and the selective emitter can be extracted (figure 4.5). The different emitter depths are clearly visible. The analysis also shows the higher dopant concentration at the top of the laser doped selective emitter compared to the regular emitter (green circle).

The ability to obtain two dimensional images of the dopant areas (in this case regular and selective emitter) with high spatial resolution, allows a

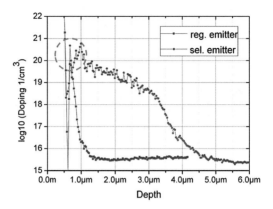

Figure 4.5: *LDSE emitter profiles.*

detailed process characterization. Different processing parameters, e.g. laser pulse energy, could be tested for their influence on the resulting dimensions of the LDSE and the dopant concentration.

4.1.2 Other Selective Emitter Concepts

Several ways for the creation of a selective emitter with improved device performance are tested by the solar cell manufacturers [117]. Beside the LDSE approach, described in the preceding subsection, the printed selective emitter technique (inkjet PSE) was characterized within this work. This process allows direct layout printing of the doping material. The ink consists of highly doped Si nano-particles. Depending on the type of ink used, the sheet resistance of the surface can be set. The small ink droplets are printed directly at the surface. The printed selective emitter (PSE) technique adds only one additional process step to an existing standard manufacturing process. The complete analysis can be found in [120].

Another approach to create different emitter types within one solar cell is the height selective emitter (HSE) concept. Although driven by the same requirements as the classical selective emitters, the doping differences in HSE apply on a much smaller scale. The selectivity is achieved within the single pyramids of a textured solar cell surface. The top of the pyramids is highly doped, whereas the bottom in between the pyramids is lowly doped; therefore the name height selective emitter. SSRM analysis could prove the dopant differences between valleys and tips of the microscopic pyramids for the HSE process in comparison to the homogeneously doped emitter of a standard solar cell [121].

4.2 Integrated Silicon Devices

In the second technology block, the application of SSRM to integrated silicon devices is described in more detail. Whereas the examples shown in the PV application section focused on process characterization for technology development, the following two analysis cases show the applicability of SSRM to local deviations in intended dopant concentration within an IC. Here, one major key for success is the large degree of freedom in positioning the cross section. Virtually any device in an IC can be prepared in any direction. Additionally, the possible feedback loop to device/process simulation data is shown for the first analysis case.

4.2.1 Resist Lift Off

To enable CMOS technology and the related realization of p- and n-MOSFETs (metal oxide semiconductor field effect transistors) on the same substrate material, wells are required. In case a p-doped substrate material is used, n-FETs do not need an extra well, but the p-FETs have to be surrounded by an n-well. As the well has to be significantly deeper than the source, drain and channel of the transistors, higher acceleration voltages have to be used during dopant implantation. Furthermore, to avoid channeling effects within the substrate and unwanted scattering effects at the resist edges, an additional tilt and twist angle of several degrees is applied to the primary beam. However, the high acceleration voltages make it necessary to use thick resist layers during implantation in order to prevent dopant punch through in actually protected areas. The thicker the resist layer gets, the higher becomes the probability of resist lift off.

An SEM image of a detached resist corner, obtained during inline inspection is shown in figure 4.6. The goal of the technology development is to avoid the resist peeling. However, in case of present resist lift off the question arises, whether there is an effect on the final position of the formed well or not.

In order to focus on the well implant, the wafer for analysis was not fully processed. No other implant then the annealed well implant is present at the test samples. The positions for SSRM analysis were determined by inline inspection, as shown in figure 4.6. The localization of the analysis sites

Figure 4.6: *SEM image of a detached resist corner, obtained during inline inspection.*

precluded cleaving for sample preparation. On the other hand the lift off areas were large enough to apply mechanical grinding and polishing instead of focused ion beam (FIB) preparation.

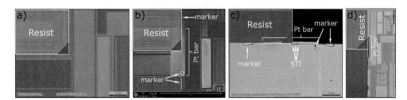

Figure 4.7: *SSRM sample preparation from the initial sample (top down OM image of the target site in a), FIB protective Pt layer deposition and marker milling b) to the final sample after grinding and polishing (cross section shown in c), top down OM in d)).*

Figure 4.7 illustrates the single preparation steps. A top down OM image of the target site a) includes schematically the position of the former detached resist corner. The specific analysis site was marked and at the same time locally protected by a FIB deposited platinum bar b). For grinding and polishing, diamond and aluminum-oxide discs were employed. For the final polishing step discs with embedded 50 nm Al_2O_3-particles were used. The resulting cross section c) exhibits the Pt protection layer, the FIB markers and the unambiguous shallow trench isolation (STI) pattern used for orientation in the later SSRM image. In 4.7 d) a top down view of the prepared sample is shown. Electrical connection was established by global scratching of the front and backside of the chip followed by the application of silver paint.

Two samples were analyzed. The first one was a reference sample with regularly attached resist and the second one the actual fail sample with detached resist at the specific analysis site. *Nanosensors* CDTP-NCHR probes were used, as the size of the features is not critically small. The applied sample bias was $V_{dc} = -50\,\mathrm{mV}$. The expected result of a resist lift off is a difference in the horizontal position of the well edge. Therefore, the scan direction was chosen parallel to the sample surface. As a positive side effect, tip wear is kept at a very low level for this scan direction, because the tip runs most of the time on the very smooth silicon surface and does not need to pass the silicon to oxide/Pt interface in every single scan line.

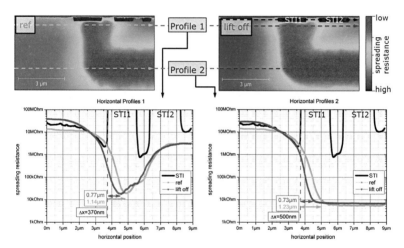

Figure 4.8: *SSRM images of the reference and fail sample are shown at the top. Horizontal profiles were extracted right below the STI areas (profile 1) and at the depth of the maximum well dopant concentration (profile 2). Comparison of the profiles clearly shows the shift of the well edge due to the resist lift off.*

The results for both samples are shown in figure 4.8. Color coding is the same. The lowly doped substrate is represented in blue and the highly doped well in red color. The maximum in dopant concentration of the well implant is in approximately $3\,\mu m$ depth. The non-conductive isolation trenches visible at the top right hand side of the individual images were used for orientation. The resist edge was located right above the outer left STI area (see figure 4.7 c). The SSRM images confirm this coarse position. A more precise determination of the horizontal position of the wells edges is possible after extracting horizontal line profiles. The bottom of figure 4.8 compares the profiles extracted at two different depths. The first one is located right beneath the silicon surface (below the STI) and the second profile cuts through the wells dopant concentration maximum. As a reference, the profile through the STIs is included in the graph. All horizontal measurements are done with respect to the most left STI edge. Both profile graphs exhibit a shift of the well edge of several hundred nanometers towards the direction of the resist lift off. This observation matches the expectation. The detached resist exposes the underlying silicon to a small degree and the dimensions of the implant are shifted into this direction.

A very graphical illustration of the effect is shown in figure 4.9. It is a difference image of the two SSRM images of figure 4.8.

$$R_{spread}(diff) = R_{spread}(liftoff) - R_{spread}(ref) \qquad (4.1)$$

Areas of comparable spreading resistance and therefore dopant concentration lead to negligibly low values in the resulting image, whereas differences become emphasized. The one obvious feature within the silicon substrate is the left edge of the well showing elevated spreading resistance values as a result of its extension in case of the resist lift off.

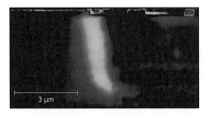

Figure 4.9: *Difference image ($R_{spread}(diff)$) of the pass and the fail sample.*

In parallel to the SSRM analysis, a process simulation using technology computer aided design (TCAD) was performed. The different simulation conditions are shown in figure 4.10. In a) a perfect resist edge is modeled,

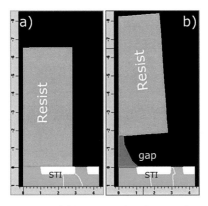

Figure 4.10: *The two TCAD setups for the simulation of the impact of a 2 μm detached resist b) in comparison to a perfect resist edge a).*

whereas b) includes a 2 μm gap between the chip surface and the resist. Simulation case b) represents quite well the condition of the detached resist.

The simulation results are shown in figure 4.11 a) and b). Although no line profiles for precise measurement are included here, the different location of the well edge is clearly visible. The differences between the two simulation cases are in good agreement with the differences between the two SSRM images. Both methods show a shift of the well edge towards the resist in case of the resist lift off.

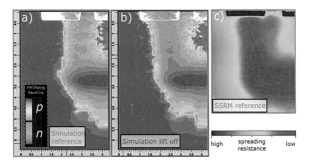

Figure 4.11: *TCAD simulations of a perfect resist edge a) and in case of 2 μm resist lift off b) in comparison to the SSRM measurement c).*

A comparison of the simulation result for the perfect resist edge a) and the

SSRM image of the reference sample (fig. 4.11 c)) exhibits a mismatch of the shape at the upper part of the wells edge. This leads to the conclusion, that the TCAD simulation is not perfect and its results differ from the actual hardware. The obtained information could be used to adjust the simulation or mask edge position.

SSRM with its high spatial resolution allows a precise determination of the position of an implant area with respect to some reference (in our case the STI edge). This allows the comparison of samples with different implant conditions. The shown analysis of a normal and detached resist edge is just one example. Data processing capabilities like 2D line profile extraction and generation of difference images are powerful features for measurement and visualization of differences in dopant area position.

TCAD based process simulations confirm the observed differences between the two samples analyzed by SSRM. On the other hand, minor differences between the simulation and the experimental data were observed. If necessary, the TCAD simulator could be adjusted to match the SSRM observations.

4.2.2 Gate Depletion

This topic will be discussed in more detail, as it consists of two separate analysis cases and a preliminary experiment. Moreover, the gate depletion analysis demonstrates the advantages of the SSRM method and represents a focus of this work: two dimensional imaging with high spatial resolution and high dynamic range applied at site specific devices as identified within the failure analysis flow. Both, the well-controlled assessment experiment and the two case studies were published [122, 123].

CMOS technologies using doped polycrystalline silicon (poly-Si) as gate material have to take gate depletion effects into account. General studies covering the physical origin and the implications on device characteristics have been carried out in the past [124, 125, 126, 127, 128]

Gate depletion occurs due to the finite doping of the poly-Si gate material. For example, considering a metal oxide semiconductor (MOS) capacitor with a p-type substrate and an n^+-poly-Si gate, for negative gate voltages both substrate and gate are in accumulation. In this case the gate can be treated as a metal. In contrast for positive gate voltages depletion or inversion occurs at the substrate below the gate oxide and in the same manner depletion or inversion can occur in the poly-Si gate. This adds an additional gate capacitance (CG) in series with the gate oxide capacitance (COX) and the silicon channel capacitance (CS) (figure 4.12). The higher capacity leads to a higher gate resistance, a threshold voltage (V_{Th}) shift of the transistor to higher values and thus to a decreased on state current (I_{ON}). An overall decrease in device performance is the result. In order to reduce gate depletion induced device degradation, processes aim for high poly-Si doping by applying high implant doses and sufficient dopant activation.

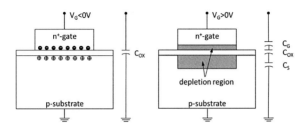

Figure 4.12: *Schematic drawing of the gate depletion effect*

Even though process technology could be adapted to suppress the negative influence of the gate depletion effect on a large scale, leading to high manufacturing yield, on a microscopic scale there are mechanisms that could lead to localized sporadic defects caused by gate depletion. The simplest one is the local implant blocking at a certain gate position, leading to a decreased gate poly-Si dopant concentration at that particular site and thus to an altered device characteristic as described above. The feasibility study and the analysis case one described in this work will engage in this topic.

The second possible reason for a failing device caused by gate depletion is the occurrence of large grains within the gate poly-Si, preventing the sufficient doping of the grain interior. Numerous work has been done, investigating the diffusion process of dopant atoms in polycrystalline silicon material [129]. The process as a whole is of great complexity and involves fast diffusion in grain boundaries, segregation between grain interior and grain boundaries, grain boundary motion and grain interior diffusion [130]. The several stages of dopant diffusion are shown in figure 4.13 under the assumption that one of the poly-Si grains is extremely large.

In the first step dopant atoms are brought into the polycrystalline silicon by ion implantation (figure 4.13a). In the following annealing step, the dopants quickly diffuse along the grain boundaries (figure 4.13b), followed by the diffusion from the grain boundaries to the grain interior (figure 4.13c). As a result of the annealing process, there is a dopant gradient from the grain boundaries to the grain interior. In case of the exceptional large poly-Si grain dopant concentration decreases so much towards the center

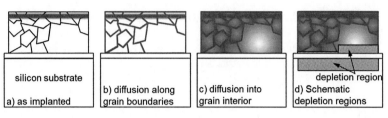

silicon substrate

a) as implanted

b) diffusion along
grain boundaries

c) diffusion into
grain interior

depletion region

d) Schematic
depletion regions

Figure 4.13: *Dopant diffusion process in poly-Si on top of a gate oxide. The red area marks the location of the dopants a) situation directly after ion implantation, b) dopant diffusion along grain boundaries, c) diffusion from grain boundaries to grain interior leading to dopant gradients and d) resulting local difference in gate depletion due to the presence of large poly-Si grains.*

of the grain, that if this big grain is located right above the gate oxide, the otherwise negligible gate depletion effect is increased and results in a degraded overall device characteristic (figure 4.13d). The second analysis case presented in this work shows an example for this failure mechanism.

In order to test the general possibility of the SSRM method to successfully analyze local regions with reduced dopant concentration and enhanced gate depletion, a well-controlled experiment was designed. Core feature of the experiment is the local selective blocking of a certain implant step during wafer processing, followed by the later analysis including localization of the artificially generated defect sites, electrical implication on integrated devices, the preparation and the analysis of the affected devices by means of SSRM.

As the test experiment has been designed to target the analysis capability of SSRM on local word line (WL) poly-Si implant blocking, one wafer, which was processed until source-drain implantation, was separated from the lot and exposed to polysterene latex (PSL) spheres. PSL spheres offer a well-defined spherical shape and they are commercially available in calibrated sizes down to diameters smaller than 100 nm. PSL spheres are usually used for testing and calibration of inline wafer inspection tools [131]. The spheres are available bottled in a solution, so the application is done by simple spraying on the wafer surface. For our experiment we have chosen spheres of 4 μm diameter. Figure 4.14 a) shows one PSL sphere located within the read only memory (ROM) array.

The so prepared wafer was processed with the regular source-drain implantation which is also responsible for word line poly-Si doping. A cleaning step followed in order to remove the PSL spheres. The wafer was then processed according to the normal process flow. Inline inspections using OM and SEM after the application of the PSL spheres, the implantation step and the removal of the spheres have been carried out in order to distinguish between process base line defects and the artificially generated defects of our experiment. The inspection also provides the defect positions for a later analysis.

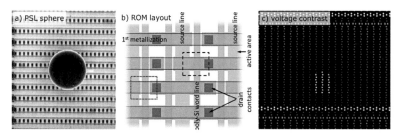

Figure 4.14: *SEM image of a polystyrene latex (PSL) sphere as applied to the wafer surface a). Layout schematic of the diffusion type ROM used b). Top down SEM image of a defect site c), showing the area of the implant blocking at contact level as spot with bright contacts within the ROM cell array.*

To be able to correlate the SSRM measurements with electrical data, the test wafer was processed until contact plug formation and the following tungsten chemical mechanical polishing (CMP) step. As a result, all integrated devices are electrically accessible via the tungsten contact level. A further processing was not necessary for our purposes as all requirements on our test sample are fulfilled: partially blocked WL poly-Si implants due to the PSL spheres, assurance to analyze the right defects and their coordinates by in line OM and SEM defect inspection and the contact level for the correlation to electrical data. Apart from that, the complete processing of the wafer would not be reasonable, since the extensive application of the PSL spheres would have led the chips to be non-functional anyway.

In order to have the same imaging conditions for a specific failing device and its unaffected reference, the analysis was restricted to defects located in the ROM of the chip. The dense ROM packing guarantees adjacent fail and reference devices, which can be imaged within one scan. The used ROM was of diffusion programmed type. For this type, bit line (BL) contacts are shared for two cells. The gate of the transistor is the word line. A common source line builds the source of the transistor. The information is encoded as the active area is printed or not. The transistors are of n-type. The principal layout of the ROM is depicted in figure 4.14 b).

For further SSRM analysis only defects located in the center of a ROM cell array with circular shape were chosen. To rule out inaccuracies of the in-line inspection coordinates, the defects were re-localized using SEM voltage contrast [132]. The approximate position of the defect was accessed using the in line defect coordinates. The spherical spot of bright white contacts within the cell array represents the artificial defect area (figure 4.14 c)). Due to the blocked implant, no source and drain diodes have formed, thus hindering the charging of the contacts during SEM imaging. The SEM voltage contrast results were confirmed by conductive AFM (c-AFM) scans and following IV-characterization of individual fail and reference contacts (figure 4.15). For both, c-AFM measurements and the IV characterization, a Multiprobe atomic force prober (AFP) with tungsten probes was utilized [133, 134].

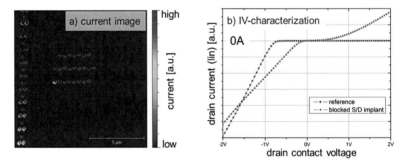

Figure 4.15: *Current image a) of a blocked implant site due to PSL application. In b) the IV-characteristics of a reference contact and a contact with blocked S/D implant is shown.*

The current image as shown in figure 4.15 a) is obtained with sample substrate at ground and the scanning probe tip at $+0.5\,\mathrm{V}$. For positive voltages all reference contacts with S/D implant are expected to show only very small leakage currents in comparison to the contacts in the area of the PSL sphere blocked implant. Both current image and IV-characteristic (figure 4.15 b) confirm the SEM voltage contrast.

For SSRM preparation, the target site was marked and covered by a platinum protection layer using FIB technique. After mechanical grinding and polishing right in front of the final cross section position, the sample was finally prepared in the FIB system.

Figure 4.16 shows the top down voltage contrast image a), SEM image of the prepared cross section b), 2D SSRM image c) and the depth profiles d) extracted from the SSRM data set of one defect generated artificially by blocking of the implant using PSL spheres.

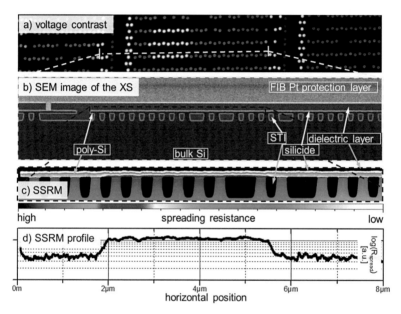

Figure 4.16: *Preparation and SSRM imaging parallel to the gate poly-Si. SEM top down image a) shows blocked implant position. The yellow dashed line indicates the position of the cross section imaged by SEM b) and SSRM c). Horizontal resistance profile through gate poly-Si is shown in d).*

In the SSRM image in c), the single active areas and the non-conductive STI can be clearly distinguished in the lower part of the image. Every visible active area represents one ROM bit with state 0. The two missing active areas on the right hand side of the image correspond to bits with state 1. The word line poly-Si lies horizontally above these features. Its completed by a highly conductive salicide. On top of that follows non-conductive inter layer dielectric (ILD). There is an explicit difference between the outer part of the word line poly-Si and the center part. By comparison of the top down

SEM image of the defect (figure 4.16) a), the FIB prepared cross section (figure 4.16) b) and the SSRM data of figure 4.16 c) it becomes clear that imaging the partially blocked word line poly-Si implant was successful and the difference in the resistance data in the center of the image is the result of a PSL sphere having blocked the implant at this position. The higher resistance center part of the WL poly-Si covers all the bits that show bright (leaky) source/drain contacts. Consequently SSRM data corresponds to the SEM voltage contrast data.

In order to support the 2D imaging results, the horizontal profile through the WL poly-Si is shown in figure 4.16 d). The profile starts at the left hand side of the SSRM scan with the beginning of the word line. The resistance value of the first two microns of the WL is constant. It is followed by a plateau with a factor of three higher resistance and approximately 3.7 μm in length. In the last part of the profile resistance values drop back to the reference level again. The observed length of the plateau corresponds well to the size of used PSL spheres (4 μm in diameter). The difference is due to the cross section position being not exactly in the middle of the circular area affected by implant blocking. The transition between high and low ohmic WL poly-Si parts is approximately 260 nm in length. This soft transition originates from the fact that the PSL spheres are in contact with the surface just in one point and thus the effect of masking the implantation is not perfect.

In a next step cross section samples perpendicular to the word line poly-Si were prepared and analyzed. Similar to the first measurement described above, the defect was first localized using SEM voltage contrast (figure 4.17 a)). Mechanical grinding and polishing up to the defect site followed. The cross section finish was done by FIB technique (figure 4.17 b)).

As a result of the rotated cross section some differences arise. The first is the reduction of analysis area. In the first example the cross section included one complete word line over several microns length. In the case of a perpendicular cross section, only small parts of different neighboring WLs can be imaged. The different WLs are separated by a dielectric. As one can derive from the SEM image (figure 4.17 c)) the ratio between dielectric and poly-Si material in the cross section is higher than 3.5. Therefore, the polycrystalline silicon area for analysis is reduced by a factor higher than 3.5.

Another drawback of this orientation is the possible presence of contacts in the cross section. In the feasibility study this could have been avoided by assigning the cross section position right between the contacts. However, in order to test for later failure analysis (FA) cases, the contacts were included in the cross section. The presence of the tungsten contacts leads to a pronounced curtaining effect during FIB cross section preparation. This effect becomes noticeable in higher overall topography and leads to imaging artifacts and a faster tip wear out.

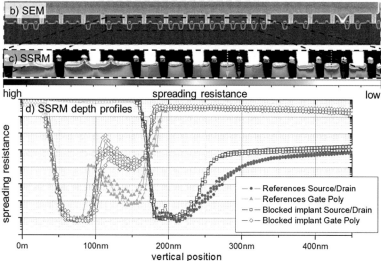

Figure 4.17: *Preparation and SSRM imaging perpendicular to the poly-Si gates. Top down voltage contrast image of the blocked implant region is shown in a). SEM image of the final cross section is shown in b) and the corresponding SSRM image in c). Vertical resistance profiles through gate conductor and STI respectively S/D region are shown in d).*

The SSRM data of the so prepared defect is shown in figure 4.17 c). All features from the SEM image in figure 4.17 b) can be recognized in the SSRM data set; very dominant the highly conductive drain contacts, in between the different word lines with the salicide on top. Drain contacts land on silicide and the source line of the transistors shows a low ohmic silicide, too.

The samples prepared parallel to the WL poly-Si were located right in the middle of the transistor channel, showing no source or drain region. This sample with the 90° rotated orientation exhibits a cross section of the whole transistor including source drain regions and the channel. Although the defect site shown in this example exhibits a large amount of STI and just a few complete transistors, a difference between the outer part of the SSRM image and the center is clearly visible.

Even more dominant is the difference in the silicon resistivity of the junction areas. As source/drain and gate poly-Si implantation is done within the same process step, this observation can also be attributed to the local implant blocking due to the PSL spheres. The SSRM image reveals the missing source-drain implant at the defect site. For this orientation of the cross section it is much harder to expose the difference in gate poly-Si doping, due to the reduced area of analysis. However, depth profiles through the gate poly-Si and the source-drain regions confirm both, the blocked implantation at the gate poly-Si and the source drain regions (figure 4.17 d)).

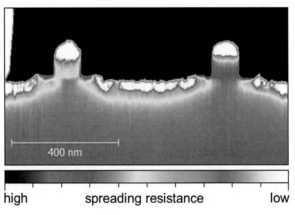

Figure 4.18: *High magnification SSRM image of the blocked implant edge. Transistor to the right shows increased gate poly-Si resistance due to blocked implant in comparison to the neighboring transistor on the left.*

As most of the defects to analyze are expected to be in the nanometer range and thus way smaller than the $4\,\mu$m defects generated using the PSL spheres, another SSRM analysis of the test defects was done using higher magnification. In order to compare reference and blocked implant sites in one image, the SSRM scan was located right at the edge of the defect region. The resulting image is shown in figure 4.18. The left hand side transistor is not affected by the implant blocking and is used as a reference. In comparison, the right hand side transistor lies within the blocked implant region. It shows a decreased local resistance for the gate poly-Si. Other details as the salicide at the top of the gate poly-Si and at the source drain regions as well as lightly doped drain implants are visible.

Analysis case 1: SRAM failure due to blocked implant

The analysis case presented in this work addresses static random access memory (SRAM) single cell fails (SCFs), which turned out to be caused by blocked word line poly-Si implant. Electrical characterization of the failing cells using an AFP showed single n-FETs of the cell exhibiting a V_{Th} shifted to higher values. This leads to smaller I_{ON} values for the affected transistor. The obtained electrical data is a reasonable explanation for the single cell to fail. Figure 4.19 shows the transfer characteristic of the failing transistor in a failing SRAM cell in comparison with a reference transistor. Red curves are linked to the logarithmic ordinate on the left and the blue curves belong to the right hand side linear ordinate. For both y-axes the solid curves are from the fail transistor and the dotted curves belong to the reference transistor. The V_{Th} is increased approximately $300\,\text{mV}$ for the fail transistor, leading to a significantly reduced I_{ON}.

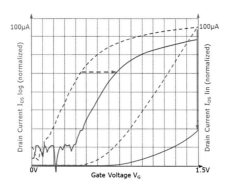

Figure 4.19: *Transfer characteristics of the failing transistor of a SCF in comparison to a reference transistor. The fail transistor shows a shifted V_{Th} to higher values and therefore a reduced I_{ON}.*

After tracking down the defect site from one failing SRAM cell to a single failing transistor, SEM and transmission electron microscopy (TEM) investigations were performed. No structural difference between the failing transistor and reference transistors could be found. In combination with the electrical data, this led to the hypothesis of an implant originated defect mechanism. One possible explanation for the observed electrical data is the occurrence of gate depletion, possibly caused by a blocked gate poly-Si implant. This hypothesis was tested by SSRM analysis.

Figure 4.20 a) shows the overview SEM image of the FIB prepared target site including a set marker for orientation. Marked within the yellow rectangle are four n-FET access transistors. The two on the left hand side of the rectangle belong to a reference SRAM cell. The two transistors on the right hand side are part of the failing SRAM cell and the one failing transistor determined by AFP is the left one of these two. As concluded from the feasibility experiments, the best cross section direction for gate poly-Si related fails is parallel to the gate conductor. Therefore, the same orientation was chosen for this analysis.

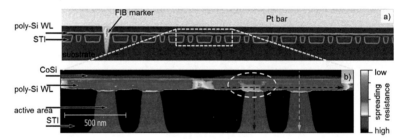

Figure 4.20: *SEM image of the FIB prepared cross section a) and the SSRM data b) of the marked failing bit and the reference bit at the left hand side. The fail transistor (second from the right) affected by the V_{TH} shift exhibits a remarkable decrease in local resistance of the gate poly-Si right above the active area.*

Figure 4.20 b) shows the corresponding SSRM image of the four access transistors of the two cells. Black areas exhibit very high spreading resistance values and correspond to dielectrics. In the bottom part of the SSRM image, the familiar pattern of the SEM image shows regions of active area with moderate spreading resistance (red) surrounded by STI, appearing in black. The highly resistive area above the word line corresponds to the ILD. On top of STI and active area lies the word line consisting of poly-Si and a salicide layer on top.

As the two transistors of the reference cell on the left hand side and the right transistor of the failing cell show no anomalies in the electric measurement, they can be used as reference. Indeed, SSRM measurement data for these three transistors are in good agreement. In comparison, the fail transistor on the left hand side of the fail cell affected by the V_{TH} shift exhibits a remarkably increase in local resistance of the gate poly-Si right above the active area. The suspicious spot covers almost the whole transistor channel at this cross section position. This result confirms the hypothesis of a blocked gate poly-Si implant causing the gate depletion.

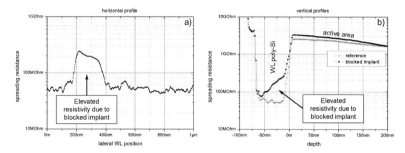

Figure 4.21: *Horizontal a) and depth b) profiles of the gate poly-Si extracted from the SSRM data shown in figure 4.20 (arrows).*

Horizontal and vertical line profiles were extracted from the measurement data (figure 4.21). The horizontal profile in a) shows the main part of the gate poly-Si exhibiting spreading resistance values around 50 MΩ. The variations visible in this reference area can be attributed to the poly-Si grain distribution in the gate. Marked in the red ellipse is the 200 nm long section with increased local resistance and thus decreased dopant concentration, responsible for the V_{Th} shift of the transistor. The depth profiles in b) obtained from the reference and the failing transistor confirm the interpretation. Whereas the reference transistor exhibits a constant dopant concentration over the whole gate poly-Si depth, the failing transistors gate poly-Si shows a decreasing dopant concentration towards the gate oxide. A little bit unexpected is the fact, that the vertical profile of the fail transistor exhibits a doping level comparable to the pass transistor near the salicide edge, despite the implant was blocked. A possible explanation of this signature would be the horizontal diffusion of dopants from the nearby implanted region.

Two artifacts visible in the SSRM image, should be explained at this point. At first, the poly-Si exhibits areas of increased local resistance in between every SRAM cell with a pair of transistors. This can be explained by the wiggled word line geometry. The poly-Si material remaining in the FIB prepared cross section at these positions with elevated resistivity is so thin, the spreading resistance is raised due to the confined geometry alone. If a provoked defect is located by random chance in such a special region of drastically thinned poly-Si, it would have been masked due to the elevated resistivity caused by the confined geometry. The second artifact is the appearance of the silicide layer with a resistivity much higher than poly-Si, which is obviously wrong, since the metallic silicide is of lower resistivity than the doped poly-Si. The most likely explanation combines the afore mentioned confined geometry present at the sample and the material removal caused by scanning under high contact forces with a very hard diamond tip in contact mode. The tip removes the remaining silicide at its thinnest positions and thereby increases the resistance of the electrical path to the back contact significantly.

Analysis case II: ROM fails due to large gate poly-Si grains

In the second analysis case single bits of a diffusion programmed ROM fail in electric test. Only ROM cells with printed active area fail, because after the intended discharging of the BL via the ROM cell transistor, the bit line is still sensed as charged. For a better understanding of the defect, failing ROM cells were IV-characterized using AFP.

Figure 4.22 a) shows the comparison of transfer characteristics of a failing and a reference cell. A shift in V_{Th} to higher values and as a result a lower I_{ON} was observed for the fail cell. The observed increase in transistor on resistance prevents a sufficient BL discharging. The nanoprobing observation fits the observed electrical fail behavior. Comparing the transfer characteristics of both gate depletion analysis cases, the compliance is evident.

For a complete characterization, structural analysis was conducted by means of SEM and TEM imaging. In contrast to the previous analysis case, a structural abnormality was found at the fail cell. As figure 4.22 b) shows, one single big poly-Si grain was observed right above the gate oxide of the failing ROM cell transistor. Taking into account the gate depletion mechanism described in the first part of the introduction, the hypothesis of a big poly-Si grain with a decreased dopant concentration as failure root cause was established. SSRM was applied to confirm that hypothesis.

The cross section of the failing ROM cell transistor is shown in figure 4.23.

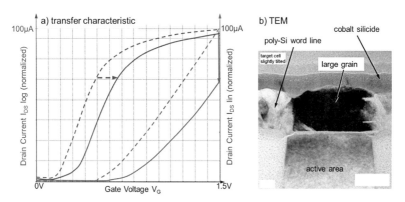

Figure 4.22: *Transfer characteristics of the failing transistor of a failing ROM cell in comparison to a reference cell a). The fail transistor shows a shifted V_{Th} to higher values and therefore a reduced I_{ON}. The corresponding TEM image of one of the failing cells is shown in b).*

The upper part of the image shows the SEM image after FIB preparation (a). Cross section direction is parallel to the word line of the ROM cell array. Substrate, transistor active area and gate poly-Si appear in dark gray. The shallow trench isolation separating the single ROM cells are imaged brighter. On top of the WL, an inter layer dielectric follows. The whole target area is covered by a FIB deposited platinum bar, ensuring electrical conductivity via contact plugs of all implant areas and acting as a protection layer.

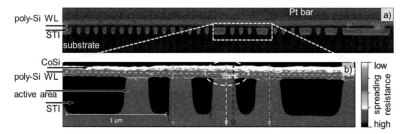

Figure 4.23: *SEM image a) of the FIB prepared cross section shown at the top and the corresponding SSRM data b) of the marked failing bit at the bottom.*

The ROM signature (1, 0, 0, 0, 0, 1, 0 in the yellow marked box) can be used for orientation, so an additional marker is not necessary. The SSRM image in figure 4.23 b) shows the same features, this time based on local resistivity. Very high resistive regions are plotted black representing dielectrics as the STI between active area and the ILD on top of the word line. Bulk substrate and transistor active area exhibit a high resistivity and appear red in the SSRM image. The metallic silicide covering the WL is very low ohmic and is drawn white. The gate poly-Si is highly doped and thus shows a low resistivity (blue), except for a small segment right at the position of the failing transistor. Starting at the right edge of the transistor up to the center of the channel, an elevated local resistivity of the gate poly-Si is visible.

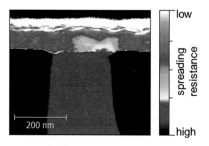

Figure 4.24: *Magnified SSRM image of the defect site revealing the detailed structure of the defect site and a dopant gradient within the affected region.*

The high spatial resolution of the SSRM method reveals the detailed structure of the microscopic defect site (figure 4.24). The big poly-Si grain and the resulting decreased dopant concentration appear to be non-uniform with a wing at each side. Within the affected region a dopant gradient is visible. Although the complete right hand side of the transistor channel is affected by decreased dopant concentration leading to elevated gate depletion, the SSRM data shows the strongest dopant decrease being located at the right corner of the transistor channel making the biggest contribution to the observed gate depletion effect.

Horizontal line scans at two different heights within the gate poly-Si are plotted in figure 4.25 a); profile 1 approximately in the middle of the poly-Si and profile 2 right above the active area level. Both profiles exhibit unusual high resistivity at the fail cell position. The decrease in dopant concentration is strongest at the bottom of the gate poly-Si (profile 2). At this position the local spreading resistance is increased by more than one order of magnitude compared to the surrounding poly-Si. In figure 4.25 b) vertical profiles of a reference position (bright green) are shown in comparison to the profiles from the fail site. The red profile is located right in the middle of the high resistive poly-Si grain; the orange profile at the edge of the big poly-Si grain. An intermediate resistance profile is drawn dark green and is located at a reference position exhibiting a slight resistance peak as shown in the horizontal profiles. The obtained SSRM data confirm the hypothesis of a single big poly-Si grain within the gate, exhibiting decreased dopant concentration and thus leading to a threshold voltage shift causing the affected transistor to fail.

The spatial resolution of the SSRM image can be estimated from the close up image shown in figure 4.24 and the profiles of figure 4.25. The full width half maximum (FWHM) of some of the features in the horizontal line scans is as small as 15 nm. In the depth profiles the transition from active area to gate poly-Si is observed to be approximately 5 nm.

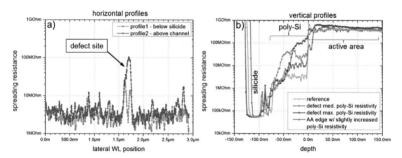

Figure 4.25: *Vertical a) and horizontal b) profiles through gate poly-Si. The vertical cut in b) compares two different profiles through the defect site to a reference profile. Additionally one profile at an active area (AA) edge exhibiting slightly increased poly-Si resistivity is shown. Both, vertical and horizontal cuts show abnormal high resistance at the fail cell position.*

A closer look at the reference area of the gate poly-Si reveals several distinct peaks in resistivity, again predominantly located right at the bottom of the gate (profile 2). These peaks coincide with the small green spots visible in the 2D-SSRM image in figure 4.23 b). The horizontal position of the peaks seems to be correlated with the active area edges. A possible explanation would be an accelerated polycrystalline silicon grain growth due to the topography at the edge positions in combination with a decreased doping density due to the bigger grain size.

In summary, a well-controlled experiment was successfully designed and conducted within this work, leading to numerous, well defined, thoroughly pre-characterized artificial defects. SSRM was applied to the so created defect sites to demonstrate the analysis capability and reproducibility as well as to test for optimum preparation and analysis parameters.

SSRM could be proved as a suitable analysis method for visualization of the generated local areas of blocked implant. The method confirms decreased dopant concentration in the gate polycrystalline silicon, which we focused on and in the source/drain implant region. Horizontal and vertical line profiles obtained from the 2D-SSRM data set reveal the effects of the implant blocking.

Furthermore two analysis cases with locally decreased dopant density of the gate poly-Si above the transistor channel as root cause for single device failure in IC environment were successfully conducted. Both applications rely on the capability of accurate on-site preparation, the high spatial resolution and the high dynamic range of the SSRM method.

In combination with complement electrical and structural analysis techniques two different mechanisms leading to the local decrease of dopant concentration are proposed. In case one, the inconspicuous structural data in conjunction with the SSRM result leads to the conclusion of a locally blocked implant during wafer processing as failure root cause. The implant blocking might be a result of resist residues, possibly caused by an uncompleted cleaning step. The second analysis case also shows a locally elevated resistance of the gate poly-Si above the fail transistor. This time the complementary structural analysis showed single big gate poly-Si grains above the failing transistor channel. The supposed failure root cause in this case is an abnormal gate poly-Si grain growth at preferred topographic features, with the effect of insufficient dopant diffusion into the grain.

High spatial resolution and the high dynamic range in combination with being not limited to test structures makes scanning spreading resistance microscopy a very powerful tool for failure analysis of integrated devices.

4.3 Power Devices

The awareness of the climate change, increasing energy prizes and the continuing trend towards more electrical systems make it necessary to efficiently process electrical energy. Power devices offer the potential for the increase in energy efficiency. This section demonstrates the applicability of

SSRM to this specific semiconductor technology. Strengths and weaknesses of the methodology are discussed. Both application cases in this section feature specialized discrete devices. Nevertheless, integrated power devices, e.g. in combination with logic circuitry, are available and have been investigated within this work. An example can be found in the literature [135].

The goal of power devices is to conduct high currents in on state (low on state resistance (R_{ON})) at small device areas (A) and simultaneously allow high breakdown or blocking voltages (V_{BR}). The product of R_{ON} times A is a characteristic of each power technology. Goal of the device developers is to minimize the area specific on resistance for a given breakdown voltage class.

The first analysis presented here, shows the column structure and the edge termination of a super-junction (SJ) metal oxide semiconductor field effect transistor (MOSFET) [98]. Due to its compensation principle it allows to achieve remarkable electrical properties. The technology characteristic product of R_{ON} times the device total area (A) at a given V_{BR} could be reduced for SJ devices below the physical limit for conventional silicon power devices.

For semiconductor devices in general, but also power devices in particular, silicon is by far the most commonly used material. Nevertheless, efforts are made to employ other materials for application in power devices applications due to their superior electronic properties, e.g. higher mobility or increased critical electrical field strength (E_{crit}). Silicon carbide (SiC) is one of those materials. The second application case demonstrates the applicability of SSRM to the SiC technology. It is shown, that some modification of the measurement setup is necessary for successful analysis.

4.3.1 Super Junction MOSFET

Super-junction (SJ) MOSFETs are widely used in power supplies or DC to DC converters. They can operate at voltages up to 1 kV, currents up to 1000 A and allow fast switching of several 10 up to several 100 kHz, depending on the voltage class and drive current. Their excellent electrical characteristics (low R_{ON} times A at high V_{BR}) are based on the compensation principle. A scheme comparing a standard power MOSFET to a SJ device is depicted in figure 4.26.

In SJ devices, just as well as in a standard power MOSFET, the current flows vertically from the front side source to the back side drain. The whole chip area is used for current transport. Under the premise of a constant chip area A, the total R_{ON} is the sum of the resistance contributions from the n^+ source, the channel, n-epi and n^+_{sub}. In the standard power MOSFET,

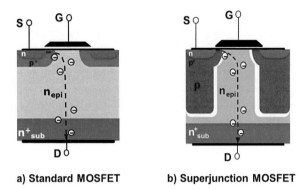

a) Standard MOSFET   **b) Superjunction MOSFET**

Figure 4.26: *Schematic drawing of a standard MOSFET a) on the left and a SJ MOSFET b) on the right hand side. In both devices current flows vertically from source to drain. The SJ MOSFET introduces p-columns within the n-epi matrix.*

(source: Infineon - Application Note - PowerMOSFETs - 600V Cool-MOS - CP - Most beneficial use of superjunction technologie devices (http://www.infineon.com/dgdl/Infineon%20-%20Application%20Note%20-%20 PowerMOSFETs%20-%20600V%20CoolMOS%E2%84%A2%20-%20CP%20-%20Most%20beneficial%20use%20 of%20superjunction%20technologie%20devices.pdf?fileId=db3a304412b407950112b40ac9a40688))

the overwhelmingly large share is the *n*-epi, as in reverse bias conditions, it has to enable the high break down voltages (V_{BR}). To do so, the *n*-epi material, which is fully depleted in reverse bias condition, needs to be doped as low as possible. Though, the lower the *n*-epi doping, the higher the resistance contribution to the overall R_{ON}. Furthermore, the critical field strength forces an increase of the *n*-epi thickness with increasing V_{BR}. Both, the low doping and the necessary thickness of the *n*-epi make it the main contribution to R_{ON}. For conventional power MOSFETs there is a limit line for marking the optimum doping profile, and therefore R_{ON} for a given voltage class [136]. The relationship between R_{ON} and V_{BR} for this limiting case is given in equation 4.2.

$$R_{on} \propto V_{BR}^{2.4..2.6} \qquad (4.2)$$

With the introduction of SJ devices this silicon limit line could be broken [137] (see figure 4.27). The main idea of the technology is to introduce alternating dopant areas of *n*- and *p*-type instead of a homogenous *n*-epi. The actual dopant concentration is chosen in a way, in the case of the transistor is in blocking state, both dopant areas get depleted equally strong. This is called the compensation principle. Dependent on the actual doping

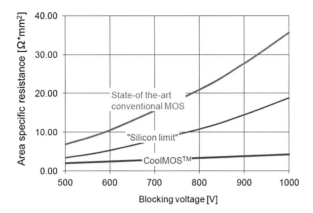

Figure 4.27: *Area specific R_{ON} versus blocking voltage of a standard power MOSFET and a SJ MOSFET. The graph is taken from an Infineon application note and therefore the SJ device is called CoolMOS, Infineons name for its SJ products.*

(source: Infineon - Application Note - PowerMOSFETs - 500V CoolMOS - CE - 500V Superjunction Mosfet for Consumer and Lighting Applications (http://www.infineon.com/dgdl/Infineon-ApplicationNote_ 500V_ CoolMOS_ CE-AN-v01_ 00-EN.pdf?fileId=db3a304336ca04c90136ea3a92e736f6))

level and the size of the alternating dopant areas, full depletion is achieved at rather low blocking voltages. Introduction of this compensation principle decouples the attainable blocking voltage of a device from the actual dopant concentration of the n-epi. It is now linked to the effective compensated carrier concentration under blocking conditions. This allows to increase the n-type dopant level of the n-epi by simultaneously increased doping of the p-areas (most commonly columns or trenches in between (figure 4.26)). As the n-epi is the major contributor of resistance in the current path of the power MOSFET, an increase in its dopant level results in a significant lowering of the devices R_{ON}. The resistance increase due to the area reduction by the p-columns is overcompensated by the resistance decrease due to higher n-epi doping. The necessary field strength to achieve the desired blocking voltage is still dependent on n-epi thickness. So the different blocking voltage classes of SJ devices are realized by deeper n-epi and accordingly deeper p-columns.

There are different ways to realize the n-epi-p-column/trench matrix. In one approach the n-epi is grown first, followed by dry etching high aspect ratio trenches that are subsequently filled with p-doped epi material. In the best case, the complete n-epi thickness can be realized by only one process cycle as described above. SSRM analyses of this kind of SJ MOSFETs exhibit the straight, high aspect ratio trenches, that are characteristic for this process technology [138, 139]. SSRM results for the second type of SJ MOSFET are shown in this work. In this approach, the n_{epi} is grown in several distinctive steps. Figure 4.28 schematically shows the p-column formation within the n-epi matrix. After each step of epi growth (black dashed lines), local discs of p-dopants are implanted. The subsequent layers of n-epi and p-discs are vertically aligned to each other. A final anneal step after n-epi completion results in an out diffusion of the implanted p-discs and therefore the formation of the aspired p-columns in the n-type epi matrix. Finally the MOS transistor is created on top of the p-columns, including a low resistive p^+-contact to the column, the source implants, contact and gate.

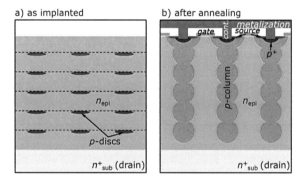

Figure 4.28: *Column formation of a SJ device. In between the subsequent n-epi growth steps, localized p-doped discs are implanted a). In a later annealing process, the p-columns are formed due to the out diffusion of the discs b).*

Since n-epi and p-columns form a matrix on a microscopic scale, application of global analysis methods like SIMS is very limited. Especially SRP cannot be applied within the devices active area, due to the alternating n- and p-type areas. Furthermore, the resulting non planar structure makes a two dimensional imaging method desirable, rather than just obtaining one dimensional profiles. In contrast to analysis of high end CMOS technology, with common device size smaller than $1\,\mu$m, analysis of SJ power MOSFETs requires large scan sizes of several ten microns in both, depth and lateral direction. For such deep structures it is hard to even perform accurate depth profiling by SIMS, due to limitations of the technique (e.g. artifacts from side wall sputtering). The commercial atomic force microscopy (AFM) used in this work offers a maximum scan size of nearly $100\,\mu$m.

However, the technology characteristics described above lead to specific requirements for the analysis of the n-epi layer including p-columns. As the common dopant concentration of both, n- and p-areas is rather low (down to values in the range of $10^{15}\,1/\text{cm}^3$), the measurement setup has to be sensitive for the resulting small signals. Increasing the applied sample bias of the SSRM measurement is one way to increase the sensitivity.

As described in subsection 3.1.2, the influence of surface states increases for decreasing dopant concentrations. Therefore a quantitative imaging of the lowly doped p-columns within the n-epi matrix demands for high quality, artifact free cross section surfaces as created by cleaving. Nevertheless, spatial properties like distortions of the p-doped trench or misalignment of the individual bubbles of the p-columns can be observed even in case surface states due to sample preparation form an inversion layer at the cross section surface.

Due to its advantages, SSRM enables for example the investigation of the p-column formation as described above. Figure 4.29 shows a SJ device at different process steps. In a) the individual epi-layers were grown interrupted by implantation of the p-discs. The subsequent annealing step leads to the p-column formation as shown in b). Figure 4.29 b) also includes the transistor of the device at the top of the image.

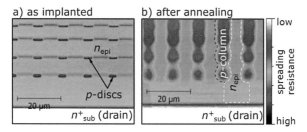

Figure 4.29: *Column formation of a SJ device observed by SSRM. In a) the individual epi layers including implanted p-discs can be observed. Annealing leads to the out diffusion of the p-discs and the formation of the p-columns.*

Obtaining two-dimensional dopant maps with high spatial resolution is of high value for the characterization of the sophisticated implant areas employed in power device technology. SSRM has the potential to support technology development, as the impact of individual process parameters on dopant area formation can be directly observed and correlated to electrical device data.

Although no FA examples are shown, the analysis cases within this work indicate the applicability of SSRM to the investigation of dopant related device failure. Vertical or horizontal misalignment of the individual parts of the p-columns for example, that would lead to a reduction in V_{BR}, could be observed directly. Also locally blocked implants can be accessed by the application of SSRM.

4.3.2 Silicon Carbide Power Devices

Silicon carbide (SiC) is a wide band gap material exhibiting superior properties for power applications in comparison to silicon. Discovered in 1824 [140] and first synthesized in 1892 [141], the first applications were of pure mechanical nature. Its extraordinary hardness (Young modulus > 400 GPa) makes SiC an outstanding abrasive. A significant increase in crystal quality after synthesis was necessary to use SiC for electronic applications.

Table 4.1: *Electrical and thermal properties of silicon, silicon carbide and gallium nitride.*

	material	Si	4H-SiC	GaN
bandgap	$[eV]$	1.12	3.26	3.39
critical el. field	$[MV/cm]$	0.3	3.3	3.3
intrinsic carrier conc.	$[1/cm^3]$	$1.4 \cdot 10^{10}$	$8.2 \cdot 10^{-9}$	$1.9 \cdot 10^{-9}$
electron sat. velocity	$[10^7 cm/s]$	1	2.2	2.5
melting point	$[1000\,K]$	1.685	3.003	2.773
thermal conductivity	$[W/cm \cdot K]$	1.5	3.8	1.3

The crystal structure of SiC is diamond cubic, similar to silicon. The unit cell consists of one center Si (C) atom and four tetrahedral bond C (Si) atoms. Different polytypes of SiC exist. For electronic applications, 4H-SiC is commonly used, due to its higher carrier mobility and wider band gap, compared to the other polytypes.

Table 4.1 shows the electrical and thermal properties of silicon, silicon carbide and gallium nitride (GaN). GaN is included, as this is another important and promising material for power applications. In figure 4.30 the properties are compared in a radar diagram. A comparison of the three material systems with respect to their application to power devices can be found here [142].

Wide band gap materials offer an increased critical electrical field strength (E_{crit}) compared to silicon. In contrast to E_{crit} values of $\approx 2.5\,\mathrm{MV/cm}$ often found in the literature, recent studies exhibit E_{crit} for SiC to be as high as for GaN ($\approx 3.3\,\mathrm{MV/cm}$) [143]. The high critical electrical field strength enables high breakdown voltages and therefore high voltage applications. Moreover, the high E_{crit} allows a tighter packaging at similar V_{BR}. Expectations are, that a discrete SiC transistor offering a V_{BR} of 1200 V would require only 10 % the size of an equivalent silicon based insulated gate bipolar transistor (IGBT) chip [144].

The intrinsic carrier concentration (n_i) is a function of the band gap. It decreases exponentially with increasing band gap. As a result SiC has a much lower intrinsic carrier concentration compared to silicon (approximately 18 orders of magnitude). This leads to a drastically decreased junction leakage at similar temperatures. Maximum operating temperature before leakage currents become dominant is around 1000 °C. In comparison, this temperature limit for operating silicon devices is approximately 250 °C [145].

SiC exhibits outstanding high thermal conductivity. The combination with the high melting point and the electronic properties as discussed above, make SiC the material of choice for high temperature applications.

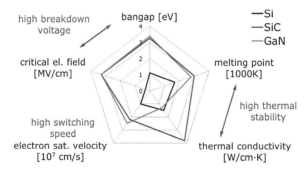

Figure 4.30: *Electrical and thermal properties of silicon carbide and gallium nitride compared in a radar diagram.*

The high electron saturation velocity (v_{sat}) guarantees low on state resistance values. Area specific R_{ON} values as low as $1\,\mathrm{m\Omega \cdot cm^2}$ seem to be achievable [146], in contrast to one order of magnitude higher $R_{ON} \cdot A$ values for best in class silicon power devices. Another result of the high v_{sat} are the high switching speeds and thus reduced switching losses. Therefore, SiC based power devices offer a higher energy efficiency. Beside the low on state resistance, the reduced switching losses are the main motivation for the application of SiC based power devices. Furthermore, high frequency applications (microwave electronics) like wireless communication and radar are enabled.

For silicon based devices, the thorough understanding of the implantation processes plays a key role for successful manufacturing. It allows very precise control of the dopant concentration and the vertical and horizontal position of the dopant atoms. A deep knowledge of damage- and annealing mechanisms has been accumulated over the years. This knowledge cannot be directly applied to SiC. Here, implantation and activation of dopant atoms requires very high annealing temperatures, as the implant process itself causes crystal damage [147].

Similar to silicon technology, silicon carbide device development requires methods for the characterization of the implantation and annealing processes. As it turns out, classical SRP profiling can only be applied in a limited way to SiC, due to the wide band gap and the exceptional hardness of the material [148]. In subsection 2.1.1, the importance of the phase transition from Si-I to the metallic Si-II for successful SRP measurement of silicon was pointed out. In SiC the first phase transition happens at much higher pressures ($\approx 100\,\text{GPa}$) [149] and it is not known to be of metallic character. This is another reason for the limited SRP applicability.

Although based on a very similar approach, SSRM provides the required sensitivity, dynamic and spatial resolution [148, 150]. The wide band gap material requires a higher dc sample bias (V_{dc}) applied between probe tip and sample (e.g. $3\,\text{V}$ [151]). In this work even higher values of 4 to $7\,\text{V}$ were used. Similar to silicon, the best analysis results for SiC were obtained at cross section surfaces prepared by sample cleaving. For fully processed, integrated devices, one disadvantage of cleaving is the presence of artifacts within the metallization, as discussed in section 3.2. The same section proposes the local leveling of the protruding metal using a hard diamond tip. This method led to good results for the SiC samples, too.

In contrast to the published work found in the literature, the SSRM measurements described within this work were carried out at fully processed, integrated devices. As a result the probe faces not only SiC, but also metal contacts or poly-Si gates within the same scan. The major resistivity differences of the different materials in combination with the high dc sample biases applied, lead to serious problems. Early experiments carried out analogous to standard SSRM showed a strong degradation of the probe tip as soon as low resistance material, as for example metallization or highly doped poly-Si, got in contact with the tip. As the samples showed normal topography, the degradation of the tip had to be of electrical nature. Hypothesis was that the probe tip could not withstand the high currents flowing at the increased V_{dc}. Therefore, higher resistance probe tips were applied. In section 3.3 (Probes) the two main probe types used in this work were discussed. The $CDTP$-type probes from *Nanosensors* are specified with 1 to 10 kΩ, whereas the $IMEC$ probes exhibit resistance values between 10 and 100 kΩ. Furthermore, probes from ADT[4] have been assessed within this worked. $ADTs$ early stage probes showed resistance values of $\approx$ 1 MΩ. Despite too high for standard silicon application, the increased overall resistance probes were tested in combination with SiC samples. Similar to the other probes, imaging of pure SiC was possible. As soon as the tip got in contact with one of the low resistance materials, an immediate dislocation of the tip could be observed. This behavior was reproducible and prevented successful SSRM measurement. The observations could be explained by fast heating and the associated bending of the tip/cantilever caused by the high dc sample biases and the low resistivity of the sample material in combination with the high resistance probes.

In order to maintain the high overall resistance of the measurement setup, but avoid degradation or heating of the probe tip itself, an additional serial resistance was introduced within the measurement path. In the experiments within this work a 1 MΩ resistor was used, integrated in the wire connecting the cantilever holder to the SSRM amplifier. By doing so, the lower resistance limit of the measurement path is moved away from the tip itself. In this configuration the conventional SSRM tips can be used. As a result, enhanced tip degradation and immediate tip dislocation can be avoided. As a drawback of the introduction of an extra resistor, the dynamic of the measurement signal is strongly limited. Due to the additional serial resistance, measured resistivity values below the 1 MΩ are hard to resolve. This modification of the measurement setup can be applied only in case the areas of interest exhibit resistivity values in the range or above the introduced resistance limit. For the SiC analysis within this work, this requirement was fulfilled.

A general example of the SSRM imaging capabilities of SiC is shown in figure 4.31. A p-doped aluminum implant within an n-epi is shown for negative ($V_{dc} = -6$ V in a)) and for positive bias voltages ($V_{dc} = +7$ V in b)) applied to the chuck. For non-disclosure reasons, a scale bar is not shown, but x- and y-axis are proportional.

---

[4]http://www.thindiamond.com/uncd-products/afm-probes/

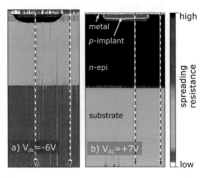

Figure 4.31: *SSRM images of a p-dopant area within an n-epi. Figure a) was obtained at negative dc bias voltages applied to the chuck (V_{dc} = −6 V), whereas in figure b) V_{dc} was set to +7 V. For non-disclosure reasons, a scale bar is not shown.*

The bottom of the image shows the 4H-SiC substrate material. A thin buffer layer between substrate and n-epi is visible at the center of the images. At the top of the n-epi layer, the p-doped area is clearly visible, surrounded by an increased n-doping area compared to the remaining n-epi.

Using the additional serial pre-resistance allows including low resistivity features like metallization or highly doped gate poly-Si within the same scan. In the shown example the metallization is clearly visible on top of the SiC material exhibiting the lowest resistivity within the whole scan.

As in silicon SSRM application, the measurement signal strongly depends on the sign of the applied voltage and the dopant type. For negative sample biases n-doped areas exhibit a lower resistivity signal with better signal-noise ratio compared to the data obtained at positive voltages. On the side, positive sample biases deliver better (lower resistivity, better S/N ratio) signals for the p-doped areas. This dc bias voltage dependency might even lead to a limitation of the measurement signal for one of the voltage polarities, due to the amplifier limitations. The effect can be seen in figure 4.31 b). Here, the n-epi appears to have superior S/N ratio, instead the signal is just limited by the SSRM amplifier module and shows its maximum value over the whole area.

The extracted profiles in figure 4.32 illustrate the observed behavior. Whereas n-doped areas show rather low spreading resistance values ($> 10\,\text{M}\Omega$) at negative voltage biases applied to the chuck in a), at positive V_{dc} as shown in b), the resistivity of the n-doped substrate is approximately $100\,\text{M}\Omega$ and

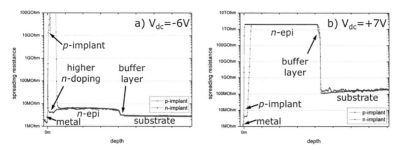

Figure 4.32: *Depth profiles extracted from the SSRM images of figure 4.31.*

the values for the n-epi represent the amplifier limit. The strong dependency of the measurement signal on the applied dc bias is of advantage to determine the local dopant type. On the other hand, the dynamic range within one type of dopant is limited by the application of an additional serial resistance. This becomes evident in figure 4.32. Spreading resistance values of the metal, n-epi and n-substrate are all within one order of magnitude, whereas the specific carrier concentrations of the three features differ by more than six orders of magnitude.

Nevertheless, the introduction of an additional, well chosen, serial pre-resistance allows two-dimensional imaging of SiC dopant areas together with low resistive material like metal or gate poly-Si within the same scan. The vertical and horizontal position of SiC dopant areas can be obtained with high spatial resolution. Differences in dopant concentration can be visualized by SSRM. Due to the strong dependency of the measurement signal on V_{dc}, predictions on the local dopant type can be made.

5 Active Device Characterization by SSRM

In the previous chapters, all experiments were carried out using one specific, constant dc sample bias (V_{dc}) applied to the device under test (DUT) as a whole and to all of its components visualized by scanning spreading resistance microscopy (SSRM) in particular. The applied V_{dc} creates a potential difference and causes current flow between the sample and the probe tip. Beside that local current flow, the device under test is in equilibrium condition under ideal SSRM operation, as all dopant areas and device components are intentionally short-circuited.

The wide acceptance of SSRM as a method for semiconductor dopant characterization proves this approach to be reasonable. In fact, the equilibrium condition of the DUT guarantees the measured local resistivity (as a function of the local free charge carrier density) to correlate with the activated dopant concentration, which is usually the desired result. Exceptions will be discussed later.

In this chapter, a motivation for the two dimensional characterization of the charge carrier density of active devices with high spatial resolution will be given. Requirements and existing approaches for active device measurement will be discussed. A test device for active device measurement is introduced (5.2). Additionally, the special requirements on sample preparation are discussed. Several scanning probe microscopy (SPM) methods are applied to the test sample in order to compare the results (5.3). Although aware of violating basic measurement requirements, standard SSRM is applied first (5.3.1), followed by the lock-in based scanning differential spreading resistance microscopy (SDSRM) (5.3.2) and scanning voltage microscopy (SVM) (5.3.3). Finally, scanning dynamic voltage spreading resistance microscopy (SDVSRM) is introduced, as a combination of SVM and SSRM with dynamically controlled and scanner synchronized sample bias voltages (5.4).

5.1 Motivation

In the beginning of this section, reasons for the introduction of a dynamic V_{dc} are given. The voltages set within the atomic force microscopy (AFM) controller software and applied to the chuck and to the device components are redefined. It is shown, that even in case of a classical back contact and the associated short circuiting of all device components, a well controlled dynamic sample bias is desirable and might be of advantage. Furthermore, the ability to apply dynamic voltage biases to the individual components of an actively driven device is one of the basic requirements for SDVSRM measurements, an enhanced SSRM method developed within this work for characterization of active devices. The second part of this motivation will discuss the benefits of the new method.

5.1.1 Dynamic Voltage

In standard SSRM, V_{dc} stands synonymously for the software set dc sample bias (V_{dc-sw}) and the dc sample bias applied to the chuck (V_{dc-hw}). Furthermore, as the DUT is short circuited to the chuck, V_{dc} is assumed to be equal to the local sample surface voltage (V_s) at any time. Moreover, V_s is supposed to be constant across the sample cross section surface. As shown later in 5.3.3 (Scanning Voltage Microscopy), despite the creation of a back contact and the associated short circuiting of the individual dopant areas, these dopant areas might exhibit significantly different potentials at the sample surface $(V_{dc-sw} = V_{dc-hw} \neq V_s)$.

In SSRM operation the scanning probe is connected to the amplifier ground potential, accordingly the probe tip voltage (V_{probe}) is defined as 0 V. As a result, V_s defines the local potential difference between probe tip and sample surface and therefore, the actual SSRM measurement voltage. As mentioned above, the assumption of V_s being equal to V_{dc-sw} is not true in every case. This leads to local deviations in the actual SSRM measurement voltage, and hence to excursions in the back calculation of the local resistivity. Especially for small values of V_{dc-sw}, the possible occurrence of deviations in V_s has to be taken into account.

A reason for the observed differences in sample surface potential are electric fields within the DUT. Internal electric fields can result from surface states at interfaces (e.g. silicon to gate oxide interface) or trapped charges in insulators. Furthermore, the incident laser light can have a significant influence on the actual V_s.

Decoupling of V_{dc-sw} and the hardware V_{dc-hw} would be a possible solution to match V_{dc-sw} and V_s again. Dynamic, scanner synchronized control of the V_{dc-hw} has the potential to improve SSRM, as it allows to constantly keep the local V_s equal to V_{dc-sw} during scanning.

5.1.2 Active Devices

Two dimensional visualization of dopant areas within semiconductor devices was a breakthrough, enabled by the application of specifically tailored SPM methods. Methods like scanning capacitance microscopy (SCM) or SSRM enabled new applications in process characterization, device simulator matching and failure analysis (FA). With the knowledge of the free carrier concentration within a device in equilibrium state, detailed predictions on its electrical behavior can be derived.

Devices and their dopant distribution are designed for electrical operation. In case of a metal oxide semiconductor field effect transistor (MOSFET) this includes switching between on- and off-state with significantly different electrical properties of the device. The field effect transistor (FET) channel resistance decreases more than six orders of magnitude from off- to on-state. Commonly used methods for active device characterization are *IV*- or *CV*-measurements, both being electrical methods showing the response of the complete device. From this integral electrical response, deductions on the dopant distribution can be made by so called inverse modeling techniques [152].

On the contrary, SSRM assesses the free charge carrier distribution of a device in equilibrium state. The obtained local data can be used to derive predictions on the integral electrical behavior of the actively driven device.

The combination of both, electrically functional devices, that can be actively driven and a simultaneous visualization and characterization of the devices free charge carrier distribution has the potential for a better understanding of the device functionality. This applies for all application areas discussed for SSRM. In device development, the simulation results might fit the actual carrier distribution for the equilibrium case, but exhibit differences under certain modes of operation. Direct assessment of the free carrier distribution in the specific mode would be of great benefit.

Another possible application case would be in failure analysis. It is perfectly conceivable to increase the fingerprint of a defect by switching from equilibrium state to some actively driven mode of operation. Physical analysis of a FET exhibiting, for example, punch through behavior in *I-V*-characterization would drastically benefit from the possibility, to set the device into off-state during analysis.

The dc biases applied to the components of the actively driven device (V_{active}) are decoupled from V_{dc-sw} and V_{dc-hw} by definition. This decoupling is neglected, in case standard SSRM is applied to actively driven devices. The SDSRM techniques strives to filter V_{active} by using lock-in amplifier technique. SDVSRM measurement principle relies on determination of V_s in a first pass, and the subsequent adaption of V_{active} in a way, the resulting V_s matches the software set V_{dc-sw} during the second pass SSRM measurement. Therefore, the dynamic dc bias applied to one component of the actively driven device ($V_{dynamic}$) depends on V_{active}, V_{dc-sw} and V_s.

All active biasing of devices as described in this work is of a static nature. A specific state of operation is set by biasing and kept constant during the whole measurement. Even in case of SDVSRM measurement, where dynamic voltages are applied, the device internal potentials stay constant during measurement. As described above, this adds new prospects to device characterization. Nevertheless, the important aspect of dynamic device behavior is neglected in this static approach. Modern microprocessor unit (MPU) devices offer switching frequencies up to several gigahertz. Although much slower, even the power devices discussed in the application chapter (4.3) exhibit significant switching speeds up to one megahertz. Effects on carrier distribution specifically related to dynamic device operation are out of range for the analysis methods, discussed here. In case of SSRM, the relatively long integration times for current amplification limit the possibility to characterize dynamic processes.

First attempts to apply the relatively new SPM techniques to actively driven devices, were made only short after SSRM was established. A method for local potential measurements was introduced in 1995 [153] and patented in 1998 [154]. This method, dubbed nano-potentiometry is equivalent to what is referred to as SVM by other groups and in this work. Results of this method applied to a MOSFET can be found in[155, 156].

In 2000, SCM was successfully applied for active device characterization for the first time [157]. As a major advantage, in SCM the probe tip is not in direct contact with the semiconductor material, thus reducing the probes electrical influence on the DUT. A second example for the application of the SCM technique includes a flexible approach to combine the sample holder and voltage application [158]. In this work, a similar setup employing packaged devices is used.

A lot of effort on characterization of active devices was put in by a group at the University of Toronto. Focus of their work was the investigation on indium phosphide (InP) heterostructure multiquantum-well lasers. Additionally to classical SCM and SSRM measurements [159], SVM was applied on a larger scale [160, 161, 162, 163]. The works culminated in the development of SDSRM, a new method focusing on the characterization of active devices [164, 165].

Both, SVM and SDSRM are discussed in more detail in the corresponding sections below. Good reviews of SVM in particular [166] and some of the methods applicable for active device characterization [167] can be found in the literature.

5.2 Test Devices

Two different samples were prepared for the active device measurements. The first one is a 65 nm transistor test array and the other an array of photodiodes (PDs). Whereas the transistor array is mainly used for demonstration of the preparation specific aspects of active device measurement, the actual SPM measurements presented in this work originate from the PD sample. The transistor test device exhibits small feature size and is sensitive to possible gate leakages occurring due to the cross section preparation. Therefore, influences from sample preparation can be assessed using this sample. On the other hand, the small feature size, especially in horizontal direction, makes an evaluation of the imaging under different driving voltage parameters (V_{active}) more difficult. For this purpose the PD array is better suited, as it offers planar implant areas of several ten microns length. This decouples changes in free carrier concentration from changes in measurement signal due to an alteration of the driving voltages of the device.

The laterally constant dopant profile of the PD sample is shown in figure 5.1. A shallow ($\approx 100..200$ nm), highly p-doped layer is present at the sample surface, followed by a moderately doped n-layer. The dopant concentration of this layer decreases in approximately 1 to $2\,\mu$m depth down to the constantly low doping level of the n-epi. The overall depth of the n-epi is 10 to $12\,\mu$m. The following substrate is of moderate p-doping. This layer stack includes two p-n-junctions. The one near the sample surface, between the highly p-doped layer and the moderately n-doped layer will form a narrow space charge region (SCR) under equilibrium conditions, whereas the deeper junction between the p-substrate and the lowly doped n-epi develops a wider SCR. Of particular interest in active device characterization are the observation of changes of the p/n-regions upon application of different bias potentials and how the dimensions of the associated SCRs are altered. The photodiode is a well suited test device, since much is known from CV characterization of this broad structures [168].

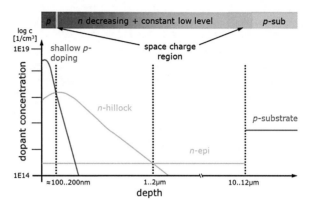

Figure 5.1: *Doping profile of the photodiode test sample. On top, there is a thin, highly p-doped layer, followed by an n-doped hillock with decreasing dopant concentration down to the low n-epi level. At the bottom the moderately p-doped substrate finishes the test layer stack.*

5.2.1 Sample Preparation

In chapter 3.2 (Sample Preparation), three different sample preparation techniques were presented; cleaving, grinding&polishing and focused ion beam (FIB) processing. In this work focus is on the grinding & polishing approach, because of several reasons. First of all, the freedom of cross section positioning should be preserved. With possible applications in FA in mind, this is a strong requirement, simultaneously ruling out sample cleaving. On the other hand, FIB with its superior accuracy, strongly limits the prepared cross section area and therefore, was not taken into account for sample preparation. By the application of grinding&polishing large cross section areas of high quality can be achieved at user specified positions with an accuracy down to 100 nm.

Another reason is, that in previous studies, it has been shown, that grinding&polishing delivers the most reliable results with respect to the minimization of surface leakage currents occurring at the samples cross section surface due to the preparation [155].

One major requirement for an active driving of several distinct components of a device is the ability to electrically contact them separately. In general, the electrical connection can be established whether before or after the cross section preparation. Though, an additional preparation step following the final cross section preparation is not recommended, due to the sensitivity of the cross section surface. Furthermore, establishing electrical connection after the various, partly dirty, preparation steps, is difficult and might not yield in low resistance contact.

Within this work, the decision was to use a wire bond approach within a test package and to start with the electrical connection of the individual device components, before cross sectioning. By doing so, the electrical contact could be achieved at the complete, non-prepared and therefore clean chip. Furthermore, this approach offers the possibility of electrical characterization, e.g. by means of IV-characterization, before, during and after sample preparation. For protection and passivation purposes, the bond wires were coated before grinding & polishing, using an epoxy resin[1]. Figure 5.2 shows the bonded sample within a package. Cutting the package in half is mandatory for the further mechanical cross sectioning of the device.

Figure 5.2: *Images of the bisected package including the DUT. Bond wires are protected by epoxy resin. The included scale is in cm.*

In order to mount the package with the included pins to the grinding tripod, a socket was glued to the tripod. The package tightly fits in the socket and is kept in position. The actual grinding & polishing was done according to the description given in section 3.2.

A part of the PD array after sample preparation is shown in the top down optical microscopy (OM) image in figure 5.3 a). One single cell is approximately 40 by $40\,\mu m^2$ in size. The center contact, which connects the shallow, top, highly p-doped layer of each cell, is clearly resolved in the image. The orientation of the cross section has to be chosen in a way, this center contact and its metal connection (dark line above) are not removed by

---

[1] *EpoThin* from *Buehler*

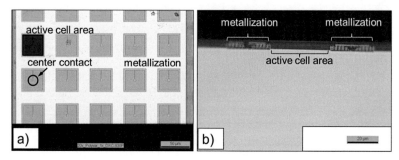

Figure 5.3: *Top down a) and cross section b) optical microscopy images of the photodiode test sample after preparation.*

the preparation process. In between the individual cells, the metallization stack is visible. A cross section image of the so prepared cross section is shown in figure 5.3 b).

5.2.2 Influence of Preparation on IV-characteristics

The prepared cross section of the transistor test array sample is shown in figure 5.4. The complete width of the transistor array is exposed with no significant artifacts due to the preparation.

Transfer characteristics of the transistor test structure is shown in figure 5.5. The *IV*-measurements were obtained before (a) and after (b) the cross section preparation. No significant differences could be observed, illustrating the maintenance of the electrical characteristics of the devices. Furthermore, no increase in gate leakage current is observed. Due to the large amount of transistors within the test device (≈ 65000) the overall gate leakage current is in the order of nano amps. The maximum increase of the leakage current, caused by the cross section preparation therefore is in the 100 pA range. Taking into account, that the cross section contains nearly one hundred individual transistors, this maximum leakage current is in the 10 pA range for a single device. This is an important fact, as it proves cross section surface leakage currents to be negligible and the electrical device characteristics to be unchanged.

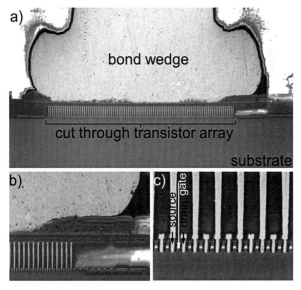

Figure 5.4: *SEM images of the prepared cross section of the* 65 nm *transistor test array. The position of the cross section was chosen to be right at the beginning of the array cutting through the center of the bond wedge. In a) the complete width of the array is visible. Images b) and c) show higher magnification images of the array edge and a close up of the individual transistor cross sections, respectively.*

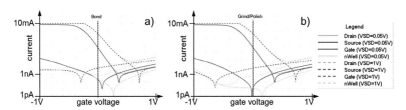

Figure 5.5: *Transfer characteristics of the transistor test array before a) and after b) cross section preparation. No significant changes could be resolved, proving the preparation process not to change electrical device characteristics and surface leakage currents to be negligible.*

5.3 Existing Methods for Active Device Characterization

5.3.1 Standard SSRM

The resistance value obtained by the *Bruker* SSRM system used within this work is calculated by dividing the V_{dc-sw} value set in the controller software, by the current measured via the SSRM amplifier. For an accurate resistance measurement, the equivalence of the software set V_{dc-sw} and the actual local sample surface voltage (V_s) has to be ensured. Obviously this requirement is violated, in case of active device measurement employing standard SSRM.

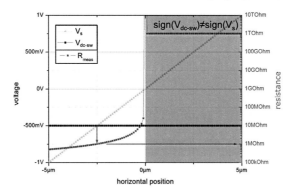

Figure 5.6: *Correlation of R_{meas} at a constant V_{dc-sw} in case V_s is increasing linear over horizontal position. In this model a sample of $1\,\mathrm{M\Omega}$ constant local spreading resistance is assumed.*

Figure 5.6 illustrates the resulting impact on measurement resistance (R_{meas}). This model assumes a sample of $1\,\mathrm{M\Omega}$ constant local spreading resistance. At both ends of the sample constant voltages are applied in a way, that V_s increases linear from -1 to $+1\,\mathrm{V}$ over the scan length of $10\,\mu\mathrm{m}$. V_{dc-sw} is set to the constant value of $-0.5\,\mathrm{V}$. Except for the position where $V_{dc-sw}=V_s$, the obtained resistance values R_{meas} differ from the real value. Even further, for positive horizontal positions when the sign of V_{dc-sw} and V_s differs, the resistance limit of the system is displayed due to the software processing. As a result, the information obtained from an actively biased sample by the application of standard SSRM is in best case not quantitative but qualitative. In worst case no information is assessable, particularly if $sign(V_{dc}) \neq sign(V_s)$.

SSRM measurement results of the photodiode sample under various bias conditions are shown in figure 5.7. The left hand side of the image shows the wafer surface and on the right, beginning at a depth of approximately $12\,\mu m$, the p-substrate is visible. During scanning perpendicular to the wafer surface, the voltage of the p-substrate (V_{p-sub}) and the n-epi (V_{n-epi}) were kept constant at $-0.1\,V$. The voltage applied to the shallow p-layer (V_p) was stepped during scanning. V_p-values are given with respect to the $-0.1\,V$. A change in V_p is recognizable by the horizontal black lines within the SSRM image. It has to be kept in mind, that the lateral dopant concentration of the device is constant, corresponding to the depth profile of figure 5.1. Therefore, all changes observed between the individual stripes are a result of the deviations in active biasing of the device. So, every measurement stripe of approximately $1\,\mu m$ thickness represents one specific electrical device condition. On purpose, the color scale is labeled "R_{meas}" and not "spreading resistance", taking into account that $V_{dc} \neq V_s$. Focus of this specific experiment was the observation of the depletion of the n-hillock and n-epi for reverse bias conditions ($V_p < V_{n-epi}$).

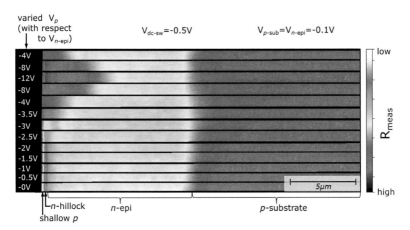

Figure 5.7: *SSRM image of the actively biased photodiode sample. V_{p-sub} and V_{n-epi} were kept constant at $-0.1\,V$. V_p was varied resulting in measurement stripes of approximately $1\,\mu m$ thickness. Although violating the rule of $V_{dc} = V_S$ reasonable quantitative effects can be observed. For example the depletion of the near surface area when increasing reverse bias condition of the upper p-n-junction.*

The corresponding line profiles are shown in figure 5.8. The $V_p = 0$ V-stripe corresponds to the sample equilibrium condition. Maximum voltage difference between n-epi and shallow p-layer is 12 V in reverse operation. The changes between these two conditions can be clearly observed near the sample surface, down to approximately 5 μm. Furthermore, the intermediate voltage states, exhibit the transition of the local carrier distribution from initial equilibrium to the deep (in terms of z-direction) depletion.

Starting from 0 V, decreasing of V_p down to -2 V shows no significant changes in carrier distribution (phase 1). Although the p-n-junction is in reverse bias direction, the n-hillock prevents an extension of the SCR into the n-epi layer. For smaller V_p a sudden decrease in carrier concentration at the n-hillock position can be observed. Once the n-doped hillock is fully depleted, the depletion region grows fast in depth with lower bias voltages V_p, as the n-epi doping is very low (phase 2). There is no change in free carrier concentration observed below the maximum depletion depth of 5 μm.

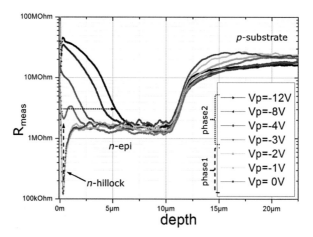

Figure 5.8: *Profiles of the individual measurement stripes with varying V_p. Strong variations within the first 5 μm below the sample surface correlating with the applied V_p bias can be observed.*

As most of the changes occur at the proximity of the wafer surface, some high magnification SSRM images of this area were obtained (see figure 5.9). Similar to the large scale scan, the depletion of the silicon can be observed for decreasing V_p with respect to V_{n-epi} and V_{p-sub}. Moreover, due to the increased lateral resolution, the shallow p-layer can be observed at the sample surface. A decrease of free carriers for higher reverse biases can be observed for this layer, too. Both, the SSRM image and the corresponding line profiles exhibit the SCR between shallow p-layer and n-hillock.

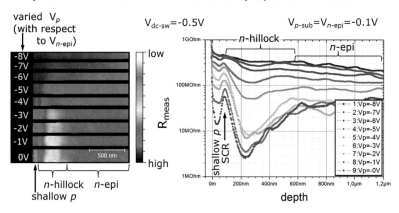

Figure 5.9: *High resolution SSRM image of the top part of the photodiode in reverse bias condition on the left hand side and and the corresponding profiles on the right. Due to the increased resolution the shallow p-layer at the wafer surface and the space charge region at the p-n-junction between shallow p and n-epi layer are visible.*

In a consecutive SSRM image, again V_p was varied, but this time positive voltage values were applied to drive the p-n-diode in forward direction. Somehow counter intuitive, the measured resistance of the shallow p-layer decreases first, before, starting from the junction position, the measurement signal jumps to the maximum value ($1\,\text{T}\Omega$). This sudden change can be explained by the software algorithm, interpreting differences in sign of the set $V_{dc-sw} = -0.5\,\text{V}$ and the measurement current (becoming positive for positive V_p values) as maximum resistance. The observed behavior is modeled in figure 5.6. This effect demonstrates one of the limitations of standard SSRM imaging applied to active devices.

Despite the significant difference between V_{dc-sw} and V_s expected within the measurement, aspects of active device behavior can be observed by the application of standard SSRM. The depletion of the first few microns of the sample under reverse bias conditions is revealed. Furthermore, several intermediate steps demonstrating the depletion process are shown. The intermediate steps reveal two stages of the depletion. In the first stage the width of the depletion region does not change, as the moderately doped

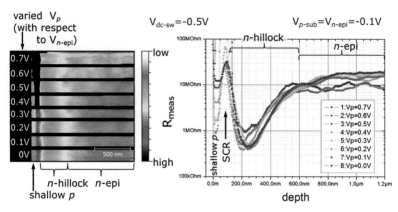

Figure 5.10: *High resolution SSRM image of the top part of the photo-diode in forward bias direction.*

n-hillock has to be depleted. Once this moderately n-doped layer is depleted, the width of the depletion region increases rapidly with increasing reverse bias, as the remaining n-epi is very low doped. High magnification images can be obtained, resolving the thin shallow p-layer and the SCR of the p-n-junction.

Application of standard SSRM to actively driven devices exhibits some serious limitations. The most drastic is the signal compliance for $sign(V_{dc-sw}) \neq sign(V_s)$ conditions as observed for biasing the diode in forward direction. A possible work around would be to decrease V_{n-epi} and V_{p-sub} further. That way, V_p could be positive with respect to V_{n-epi}, but still be negative with respect to V_{dc-sw}.

Obviously, decoupling V_s from V_{dc-sw} limits the information value to only a rough qualitative level. The later introduced SDVSRM approach strives to eliminate this decoupling.

5.3.2 Lock-in Amplifier based SDSRM

The basic idea of scanning differential spreading resistance microscopy (SDSRM) is to separate the measurement signal into two parts. Therefore, the dc bias voltage applied to the DUT is superimposed by a small amplitude ac voltage. The dc part matches the voltage and current conditions of the active device, whereas the second, ac part of the signal corresponds to the SSRM measurement conditions. The two signals can be decoupled and separately measured, theoretically avoiding a cross talk as observed using standard SSRM.

In one approach, the dc current flowing through the probe tip is filtered by a capacitor first, before the ac component of the current is fed into a lock-in amplifier and the amplifier output is fed back into the SPM system via the *Bruker* signal access module (SAM). A more detailed description of this approach including a setup scheme and application examples can be found in the literature [164].

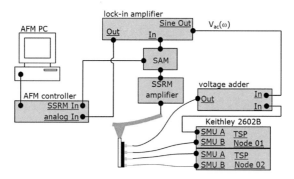

Figure 5.11: *Setup scheme of the SDSRM approach presented in this work. In contrast to the standard SSRM approach discussed in the preceding section a lock-in amplifier and a voltage adder is introduced.*

Within this work a different approach is presented, also employing a lock-in amplifier for phase sensitive detection of an ac modulated current. The setup scheme is shown in figure 5.11. The dc bias voltages (V_{active}) are generated by source measure units (SMUs) (*Keithley* 2602B) connected to the individual package pins. One of the dc voltages is superimposed by a lock-in amplifier (*Stanford Research SR830 DSP*) generated small ac bias using a voltage adder. Although in this work only one dc voltage was overlaid, in principle the setup can be extended to all dc channels. In contrast to the approach described above, the probe tip current is fed into the logarithmic SSRM amplifier first. Then, the output signal of the SSRM amplifier is split into the SSRM input-port of the AFM controller and additionally fed into the voltage input of the lock-in amplifier. Hence, the lock-in amplifier was not used for current signal amplification, but only for frequency and phase sensitive detection. The output signal of the lock-in amplifier is finally fed into an auxiliary analog input port of the AFM controller for image acquisition.

In this scheme, the signal distortion due to the application of the logarithmic SSRM current amplifier before the lock-in is a possible error source, to be considered in the discussion of the results. On the other hand, the applied scheme enhances the dynamic range for the ac-signal, due to the use of the logarithmic amplifier. The reference frequency (ω) has to be chosen to be below the bandwidth limit of the SSRM amplifier to pass through to the lock-in amplifier, but high enough to obtain a stable lock-in signal by time-integration, so the dc SSRM signal is averaged out by the lock-in amplifier for every single measurement point in the line scan. For the experiments shown here, frequencies around 1 kHz were used with an ac-amplitude of about 100 mV and the phase adjustment for the lock-in measurement was carried out on a small area of homogeneous signal before imaging. The in phase component (X) of the lock-in amplifier was recorded during image acquisition.

In principal, the introduction of an additional ac component with splitting of the effective probe tip current into a standard SSRM part and a lock-in amplified part allows various measurement configurations of different interpretation. First of all, the additional lock-in amplifier signal fed into the SPMs analog input can be obtained and presumably delivers complementary information in addition to the SSRM data channel. But, furthermore, it can be chosen between the application of the dc bias only, ac bias only, or the superposition of both to a DUT pin.

Figure 5.12 illustrates some of the possibilities based on the example of the photodiode test sample. Both, the SSRM signal and the lock-in amplifier signal are shown for two different electrical configurations. Focus of this experiment is the careful observation of the two stages of depletion within the n-doped layers. For that purpose, the connection of n-epi and the p-substrate were short-circuited, so the photodiode is operated as a two-terminal device with ports to the shallow p-region and the n-epi (p-substrate). Two different configuration of voltage supply were examined. In both cases, the dc bias applied to the shallow p-layer was superimposed by the ac voltage component. However, in configuration a) only the pure ac-voltage component was put on the n-epi terminal with no dc bias, whereas in configuration b) only a small dc bias without any ac-voltage component was used.

As in the preceding experiments, the dc bias component of the shallow p-layer (V_p) was varied during image acquisition, resulting in individual stripes of constant V_p. Some specific values are marked within the images.

On the left hand side of figure 5.12, SSRM and lock-in amplifier data for the condition of ac bias component only applied to p-sub and n-epi is shown (configuration a). Both images show the sudden decrease of the n-hillock free charge carrier distribution for V_p values below -2 V. The SSRM signal maps the shallow p-region only for $V_p = 0$ V. Below this value the SSRM amplifier output is in compliance for the shallow p layer, the adjacent SCR and, after depletion of the n-hillock, also for the upper part of the n-doped region. This behavior is not understood, as V_{dc-sw} was set to a negative value (-0.5 V) and therefore, signs of both V_{dc-sw} and the resulting probe current should be the same. More important, the lock-in signal for configuration a) shows very nicely the expected behavior of the n-doped region with increasing reverse bias voltage at the shallow p-layer. For low reverse bias, the n-hillock is not depleted and a high signal due to the increased doping with respect to the n-epi region is observed. Especially for the high reverse bias (-3 V), a homogenous, very low signal below the signal of the non-depleted n-epi layer is obtained for the entire region of the depletion, which extends far beyond the n-hillock into the n-epi layer. However, the lock-in signal from the shallow p-layer is much smaller than expected. Since the phase adjustment was done on the n-epi side, the reduced signal in the X-component of the lock-in can be caused by a phase-shift across the upper p-n-junction and not only by a reduced signal amplitude of the current response. Nevertheless, the weaker signal of the shallow p-layer still exhibits some changes in the expected trend with the variation of the dc bias, e.g. it decreases for increasing reverse bias and the slightly darker gap between shallow p and n-hillock might be interpreted as SCR.

In configuration b), where the ac-signal is only added on the dc bias of the shallow p-layer, the SSRM signal corresponds quite well to the image obtained in the case without the lock-in technique, except for deviations on the shallow p-layer. But, it is the lock-in amplifier signal of configuration b), that shows interesting different properties. As both, n-epi and p-sub are supplied with the dc bias only, their contribution to the free carrier concentration is filtered out by the lock-in amplifier. This is apparent for V_p values between $+0.6$ and approximately -1.8 V, as the lock-in output values for the n-doped layers becomes virtually zero. Only the uppermost part of the n-hillock seems to change carrier concentration with the reference frequency ω, probably due to the capacitive coupling of the ac driven shallow p-layer through the SCR.

The diode forward voltage is expected to be exceeded at V_p values higher than 0.7 V with respect to V_{n-epi}. Fore those V_p values the diode is in forward direction and a significantly high current flows. Therefore, the ac

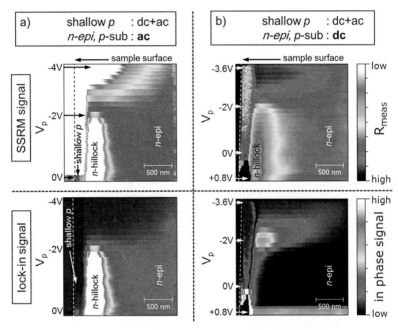

Figure 5.12: *SSRM and lock-in amplifier signal for two different electrical configurations. In all images the dc bias component of V_p was varied. While in a) n-epi and p-sub are connected to the ac signal, in b) both are connected to a dc type signal. The obtained lock-in amplifier signal is the in phase component (X) of the measurement signal.*

component applied to the shallow p-layer is also affecting the n-doped hillock and underlying epi. This effect becomes qualitatively visible within the lock-in signal of configuration b).

As the lock-in amplifier filters all dc components of the local free carrier concentration in configuration b), the depletion of the n-layers shows a different signature in comparison to the other three images. For $V_p > -2\,\mathrm{V}$ the ac coupling through the SCR is limited to the top of the n-hillock. Moreover, the induced change in carrier concentration appears to be low. For $V_p < -2\,\mathrm{V}$ the moderately doped n-hillock is depleted and ac bias therefore effects deeper n-regions. As a counter effect, the depletion itself decreases free carrier concentration in these areas. The resulting image shows the convolution of the effective ac bias and the local free carrier concentration. This way, the observed maxima around $V_p = -2\,\mathrm{V}$ can be explained.

The four different signals shown in figure 5.12 were further evaluated and compared with respect to the n-hillock depletion. Therefore, profiles through the hillock center position along the V_p-axis were extracted (figure 5.13). All profiles were normalized in order to compare the transition from moderate free carrier concentration at equilibrium conditions to the depleted state. Although, the observed sudden depletion of the n-hillock is visible in all four profiles, there are significant differences in the slope of the transition. Both signals obtained under condition a) exhibit steep slopes with a signal drop from $90\,\%$ down to $10\,\%$ between $V_p = -1.6\,\mathrm{V}$ and $-2.4\,\mathrm{V}$. For condition b) the slope within the SSRM image is significantly shallower. The overlay of both, the SSRM and lock-in signal of condition b) shows very good agreement for $V_p \le -2\,\mathrm{V}$. This supports the assumption of the lock-in signal in this case being a result of the local carrier concentration convoluted with the position, the ac bias is effective to. Both, effective ac bias area of influence and local carrier concentration show a dependency on V_p.

Additional CV-measurements indicate a full depletion of the n-hillock at $V_p = -2.5\,\mathrm{V}$ [168]. Especially, in case the ac bias is applied to all DUT components, as in condition a) shown here, the full depletion is observed for V_p values below $-2.4\,\mathrm{V}$, which corresponds very well to the CV results.

Lock-in assisted SSRM or SDSRM offers the possibility to acquire complementary information on the DUT in addition to standard SSRM. The possibility to apply dc-, ac-biases or a superimposed signal of both to the individual device components and the acquisition of the lock-in amplitude, additionally to the SSRM signal, significantly increase data acquisition possibilities. For the study of the n-hillock depletion within our photodiode test sample, it was demonstrated that the application of the ac reference bias

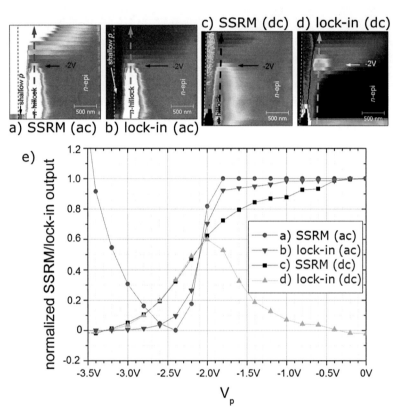

Figure 5.13: *SSRM and lock-in amplifier signal for the case the ac bias was applied to n-epi and p-sub are shown in a) and b). The images of the two channels for the case n-epi and p-sub are connected to the dc voltage are shown in c) and d). The corresponding line profiles through the center of the n-hillock were normalized and are shown in e). The steepest slope of the n-hillock depletion can be observed when the ac bias is applied to both, shallow p and n-epi.*

to the affected dopant areas improves sharpness of the observed transition significantly.

In terms of actual carrier concentration, the method, as applied in this work, is only of qualitative nature. Several of the observed effects, e.g. signal compliance, are not yet fully understood. One possible error source is the distortion of the measurement signal by the logarithmic SSRM amplifier before the phase sensitive signal detection within the lock-in amplifier. In addition, phase changes of the ac signal as potentially introduced by different signal routing to the terminals or by device internal capacities could lead to artifacts, too. Since both issues could not be circumvented in a simple way, the lock-in approach was not further investigated and improved, but an alternative method using SVM was developed. It should be added from the obtained data for actively driven devices, that the employment of a lock-in amplifier could be of benefit even for characterization of DUTs under equilibrium conditions.

5.3.3 Scanning Voltage Microscopy

SVM was the first known method to be used for the two-dimensional characterization of dopant areas of actively driven devices [153]. By mapping V_s, SVM delivers complementary information to other analysis techniques, e.g. SSRM or SCM. That alone, makes it an interesting technique worth discussing within the scope of this work. Furthermore, application of SVM is a key feature of the SDVSRM technique introduced within this work (5.4). It acquires actual V_s, enabling an SSRM measurement not based on the software set, static V_{dc-sw}, but on the actual voltage, obtained *in situ* at the sample surface.

Due to V_s not being equal to V_{dc-sw} under certain circumstances, the application of SVM can yield valuable information for SSRM of devices at equilibrium conditions, too.

In SVM the probe tip is connected to a voltmeter instead of a current amplifier as used in SSRM operation. The current flowing through the probe is maintained zero by adjusting the probe tip voltage at every measurement point to match the sample surface voltage (V_s). As a result, SVM maps the local surface voltage of the DUT. The SVM measurement setup is shown in figure 5.14.

Requirements for the employed voltmeter are a high impedance, to avoid influences of the probe tip to the sample surface potential and a high bandwidth capable of data acquisition at typical SPM scan frequencies. In this work, a *Keithley electrometer* type 617 was used. The bandwidth

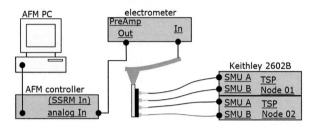

Figure 5.14: *Setup scheme of the SVM method as used within this work. The probe is connected to a voltmeter measuring the sample surface voltage.*

of the tool was tested by applying a sinusoidal signal of 2 V amplitude generated by an arbitrary wave form generator to the electrometer input. The device output was monitored by an oscilloscope. There are two analog outputs available, that can be directly fed in the auxiliary analog input of the AFM control box. Whereas the "2V Analog Output" converts the electrometer measurement signal into a ± 2 V range, the pre-amplifier output "PreAmp Out"can output up to 250 V, requiring special precaution if used as controller input for SVM. The "2V Analog Output" exhibits a bandwidth limitation of approximately 10 Hz. This is well below the required speed for data acquisition. In a first approximation, using a slow scan rate of 0.1 Hz per line and a poor resolution of 128 pixels, the minimum electrometer output bandwidth has to be around 25 Hz. Using more common imaging parameters , e.g. scan rate of 0.5 Hz per line at 512 pixels resolution, the required electrometer bandwidth is approximately 500 Hz. The pre-amplifier output, offering a cut-off frequency of about 1 kHz, fulfills the bandwidth requirement. Therefore, it is used for SVM data acquisition.

SVM measurement capabilities were tested on the photodiode sample, described earlier. Line profiles extracted from high magnification images of the top shallow p-layer and the underlying n-region, are shown in figure 5.15. During image acquisition V_{n-epi} and V_{p-sub} were kept constant at 0 V, while V_p was varied. On the left hand side in 5.15 a), the sensitivity of the method was assessed. Differences in V_s between the individual dopant layers as small as 2 mV were successfully resolved. Additionally, an offset of -11 mV could be detected in this high accuracy measurement. In 5.15 b) more common voltages were applied to the shallow p-layer. For reverse bias conditions of the photodiode, the measured V_s at the shallow p-layer position is equivalent to the applied V_p. A different, yet also predictable, behavior is observed for the diode forward direction bias conditions. For $V_p \geq 0.5\,\text{V}$ significantly smaller V_S values are measured at the corresponding p-layer position. Moreover, for higher V_p values, a limiting V_s of approximately $0.6\,\text{V}$ was observed at the shallow p-layer, by simultaneous increase in V_s of the underlying n-layer. Another subtle detail is revealed, when looking at the extension of the observed V_s signal in depth dependent on the applied V_p. As for negative V_p values down to $-0.5\,\text{V}$, the position, where V_s declines to $0\,\text{V}$ stays constant, for positive values of V_p, the observed depth of elevated V_s increases. Again, this effect can be attributed to the different device states, i.e. reverse and forward bias condition of the p-n-diode.

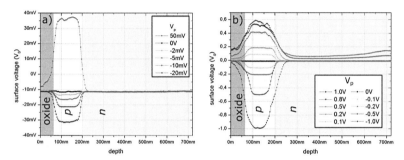

Figure 5.15: *SVM line profiles extracted from high magnification scans. In a) the sensitivity of the method is assessed. Differences in V_s as small as $2\,\text{mV}$ were resolved. In b) the different electrical behavior of the p-n-junction is observed, depending on whether the diode is biased in reverse or forward direction.*

The indicated broad minimum in V_s right below the shallow p-layer, is examined in more detail in figure 5.16a). The bigger scan size and the associated deeper profiles, prove the existence of a local minimum in obtained surface voltage correlating with the position of the moderately doped n-hillock. The observation can not be explained by considering the semiconductor device on its own. Most probably the local minimum in surface voltage is the result of an comparatively low contact resistance of the metallization to the n-hillock layer.

Furthermore, in this graph the increase in V_{n-epi} for increasing V_p, caused by the diode being biased in forward direction, becomes clearly visible. The low resistive connection between p- and n-doped layers in this state is also the reason for the observed V_p compliance at approximately $+0.6\,\text{V}$.

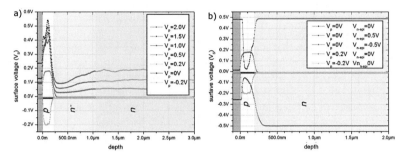

Figure 5.16: *Large depth SVM line profiles of the photodiode test structure. In figure a) V_p was stepped from -0.2 to $+2$ V, while V_{n-epi} was kept constant at 0 V. It shows the increase in V_{n-epi} for increasing V_p and the associated dip in V_s at the n-hillock position for this forward biased diode. In b) V_{n-epi} was additionally varied during image acquisition. The actual surface potential is observed correctly by the SVM technique.*

In another experiment at the photodiode test structure, in addition to V_p, the n-epi voltage (V_{n-epi}) was varied. The results are shown in figure 5.16b). It demonstrates the benefits of the SVM method. Even for $V_{n-epi} \neq 0$, the actual surface voltage at the shallow p-layer and its potential difference to V_{n-epi} can be obtained correctly.

Beside the photodiode test structure, a super-junction (SJ) device in equilibrium conditions was imaged by means of SVM. In dark mode, two images were obtained during one scan (see figure 5.17). The only difference between the two images is the laser light that is applied at the surface in a) and is switched off during scanning in b). The different scale bars for both images have to be noted. This experiment corresponds to the one shown in chapter 3, in the paragraph "light induced carrier generation".

The extracted horizontal line profiles are shown at the right hand side of figure 5.17. First of all, a small (≈ 10 mV) offset is observed, but can be neglected for further discussion. The second, more important information within the profiles, is the occurrence of local surface voltage differences. The p-doped bubbles forming the p-columns, exhibit a higher V_s value compared to the n-type epi for both illumination conditions. The third observation reveals significantly larger V_s differences between p- and n-areas in case laser light is applied to the surface ($\Delta V_s \approx 130$ mV), in comparison to the case, where the laser is switched off during scanning ($\Delta V_S \approx 10$ mV). A 10 mV difference from the set V_{dc-sw} (typically between 50 and 500 mV) in standard SSRM has only a small effect on the obtained resistance values (R_{meas}). In contrast, the observed 130 mV difference in V_s, in case laser light is applied to the sample, cannot be neglected anymore. Application of the new findings to the SSRM results shown in chapter 3 ('light induced carrier generation") leads to an enhanced understanding of the effect of light on SSRM measurements of super-junction (SJ) device columns. The carriers generated by the laser light are separated by the p-n-junction. The electrons drifting into the n-doped area can freely distribute within the n-epi. On the other hand, the generated holes are accumulated within the p-column. The only conduction path leads through the high resistive column to the top of the sample. This high resistance is further increased by the cross sectioning of the device. In summary, the possible explanation for the observed decrease in spreading resistance within the p-columns is a combination of the increase in local carrier concentration by the accumulation of holes and, additionally, the resulting increase in local surface voltage at a constant V_{dc-sw} .

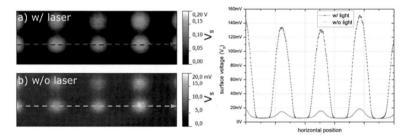

Figure 5.17: *SVM images of a part of a super junction device. During measurement in dark mode, two images were obtained, one with the laser switched on a) and one with the laser off b). Please note the different scale bars for the two images. V_s differs more than 120 mV between p-columns and n-epi if laser light is applied to the surface.*

SVM is a straight forward SPM method for the two dimensional mapping of surface potentials (V_s). It delivers complementary information for analysis of actively driven devices, in addition to the data from free carrier mapping techniques, like SSRM. Furthermore, SVM revealed existing local differences in V_s, even for a DUT in equilibrium conditions as used in standard SSRM. Therefore, complementary SVM results have the potential, to improve standard SSRM data evaluation.

5.4 Scanning Dynamic Voltage Spreading Resistance Microscopy

5.4.1 Measurement Principle

SDVSRM is a technique, combining SVM and SSRM. It is introduced for the first time within this work. The basic idea of the new technique is to map the local sample surface voltage in a first pass, followed by the actual SSRM measurement, making use of the obtained local V_s values. For that matter, the SSRM measurement voltage (V_{dc-sw}) is added to the before acquired V_s map. As a result V_{dc-sw} is not decoupled anymore from V_s. In case of present V_s differences across the sample surface, the resulting sample bias during SSRM data acquisition ($V_{dynamic}$) becomes dynamic, therefore scanning *dynamic voltage* spreading resistance microscopy (SDVSRM).

A theoretical example, illustrating the dynamic voltage measurement principle and the relationship between the involved voltages is shown in figure 5.18. Assumed is a thin sample of p- or n-type doping exhibiting a front- and back side contact. $V_{active1} = +0.5\,\mathrm{V}$ is applied at the front, and $V_{active2} = +2.5\,\mathrm{V}$ at the back side. The potential difference between front- and backside, thereby is $\Delta V_{active} = 2\,\mathrm{V}$. A decreasing dopant concentration from the front- to the back side is assumed, for no other reason than to introduce a non linear voltage profile V_s as observed by SVM in the first pass of the SDVSRM measurement. As the probe voltage ($V_{probe} = 0\,\mathrm{V}$) is kept at the amplifier virtual ground potential at any time, the whole sample potential has to be adapted during the second pass SSRM measurement in a way, that first of all the potential difference between sample front- and backside stays constant during measurement ($\Delta V_{active} = 2\,\mathrm{V} = \Delta V_{dynamic}$) and second of all, the potential difference between the probe tip (V_{probe}) and the local surface voltage ($V_s(x)$) is kept constant and equal to the SSRM measurement voltage (V_{dc-sw}). Therefore, the device bias voltages have to be dynamically controlled during SSRM data acquisition ($V_{dynamic}(x)$). The relationship is shown in the following equation for terminal number i:

$$V_{dynamic}^i(x) = V_{active}^i - V_s(x) + V_{dc-sw} \tag{5.1}$$

For the verification of the equation, the positions of front- and back side can be considered. For both, the according V_{active} equals the SVM measured V_s at these positions. Therefore, $V_{dynamic}$ of the respective device component becomes V_{dc-sw}, the desired sample bias with respect to the probe voltage (V_s).

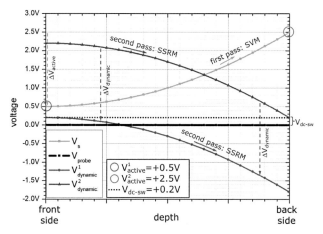

Figure 5.18: *Example plot demonstrating the correlation of the different voltages to be considered in SDVSRM. Assumed is a potential profile from the front- to the back-side of the sample (V_s - green curve). This profile is defined by the active voltages (V_{active}) applied to the front- and back-side of the sample. The V_s-profile is measured in the first pass (SVM) of the SDVSRM measurement. As the probe tip is at every time at ground potential (V_{probe} - black curve), the different device terminals have to be dynamically driven ($V_{dynamic}$) during second pass SSRM data acquisition, dependent on the probe tip position (depth). This is mandatory in order to keep the intended SSRM measurement bias (V_{dc-sw}) constant at every scan position by simultaneously maintaining the potential difference (ΔV_{active}) between the sample front- and back-side.*

5.4.2 Experimental Setup

The setup developed within this work is shown in figure 5.19. The included SVM measurement requires an electrometer. The following SSRM measurement makes use of the SSRM amplifier. To be able to switch between both, a relay was included in the setup. Switching has to be synchronized with data acquisition. Therefore, a relay capable of being triggered by a transistor transistor logic (TTL) pulse was used[2]. TTL based latching option was thought to be required, for switching based on the TTL end of line (EOL) or end of frame (EOF) pulses, generated by the AFM controller. The setup, finally realized, does not rely on the latching option.

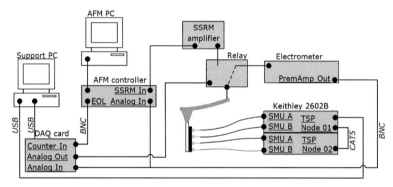

Figure 5.19: *Setup scheme of the SDVSRM technique.*

There are three possible synchronization options: per pixel, per line and per frame. On the one hand, per pixel synchronization places too high demands on relay switching speeds. On the other hand, synchronization per frame excessively increases the time between V_s measurement and the spreading resistance measurement, based on the $V_s(x)$ input. Artifacts caused by sample drift or tip wear would become more likely. Moreover, the length of the feedback loop for adjusting synchronization or imaging setting would be unacceptable. Based on this considerations, the per line synchronization was realized. In addition to the standard SSRM signal routing, the electrometer pre-amplifier output is fed into the AFMs controller analog input and simultaneously in a data acquisition (DAQ) device. The

---

[2]Customizable relay board with latching option and active high/low option from *Engineeringshock - http : //www.engineeringshock.com/custom − relay − board.html*

employed DAQ card features digital, as well as analog in- and outputs [3]. The acquired electrometer output data (V_s) is processed by a support computer, adding the SSRM measurement voltage (V_{dc-sw}). The DAQ card also counts the EOL pulses of the AFM and applies the required switching signal, based on the number of EOL pulses, to the relay trigger input.

For DAQ card input readout and output control, the *LabView* visual programming environment was used. As the employed SMUs provide *LabView* driver software, they could be controlled directly throughout the software via universal serial bus (USB) connection. During SVM operation, the individual SMU outputs are set to the constant dc voltage, defining the devices active state V_{active}. After switching to SSRM mode, the afore acquired local surface voltage (V_s) plus the SSRM measurement voltage (V_{dc-sw}) is subtracted from the individual V_{active} and then applied to the SMUs. The resulting $V_{dynamic}$ are dynamically applied, synchronized with the scanner position.

Beside some static tools to manually trigger the relay, initialize and set the SMU output and to test the single program functions, a *LabView* program was developed for active control of the SDVSRM measurement. The block diagram of the measurement control software is shown in figure 5.20. It consists of individual blocks, included in a while loop. The while loop runs, as long the SDVSRM measurement is ongoing. If the software is started, block 1 sets the voltages of the individual device components (V_{active}). Block 2 reads out the DAQ analog input card values from the electrometer.

The blocks labeled 3 read out the digital counter of the DAQ card, connected to the AFM EOL TTL signal. Whenever, the scanner changes its fast scan direction, i.e. trace-retrace switchover, the counter is incremented by one. Two counters are necessary, one for the first SVM and one for the second SSRM pass.

The blocks 4 trigger the relay by application of whether low or high TTL voltage level to switch between first pass SVM and second pass SSRM back and forth.

The calculation of $V_{dynamic}$ is divided into two parts. The first one, block 5, adds the SSRM measurement voltage (V_{dc-sw}) to the before acquired vector of V_s values. The second one is placed in the second while loop, consisting of the two blocks 3 and 6. For each loop iteration, one value of the $V_s + V_{dc-sw}$ matrix is read out, starting from the end of the matrix towards the beginning. The reversal of the sequence, is caused by the scan direction reversal from first to second pass. Subsequently, the $V_{dynamic}$ is calculated

---

[3]*NI myDAQ* from *National Instruments* - $http://www.ni.com/mydaq/$

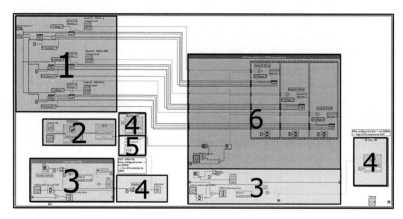

Figure 5.20: *Block diagram of the developed LabView program. A description of the individual blocks can be found in the text.*

from V_{active} values from block 1, minus the latest element extracted from the $V_s + V_{dc-sw}$ matrix.

Finally, in block 6 the $V_{dynamic}$ values for SSRM measurement are applied to the individual SMUs. In every loop iteration a new $V_{dynamic}x$ value is applied to the corresponding scanner position. After the next EOL TTL pulse, the second pass SSRM loop is stopped, the relay is switched back to SVM input (block 4) and the cycle starts over.

Minimum cycle time for the while loops was found to be 4 ms, limited by the SMU response time. Thereby, maximum scan rate for a certain resolution (points per line) is limited, too. 4 ms per point of a 512 points per line scan, lead to a minimum scan time of approximately 2 s per line, which corresponds to a scan rate of 0.25 Hz.

The program also includes a timing parameter, accounting for the delays occurring in the DAQ process and data output. The correct settings for signal synchronization are tested by simultaneous data acquisition of the SVM and $V_{dynamic}$ SMU output signal, that is redirected to the AFM controller input for that purpose. When both signals match in fast scan direction, as shown in figure 5.21, the appropriate timing parameters are found. The two SVM profiles show the top of the photodiode test sample. Label a) corresponds to $V_{active}^1 = V_p = +0.3\,\mathrm{V}$ and $V_{active}^2 = V_{n-epi} = V_{n-sub} = 0\,\mathrm{V}$. Settings for the measurement with label b) are the same for V_{active}^2, but V_{active}^1 was set to $+0.2\,\mathrm{V}$. The obtained SVM data is recorded by the DAQ card and used

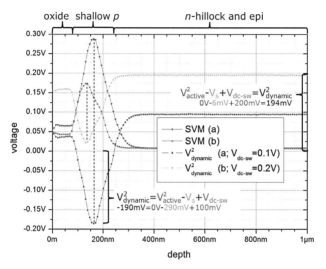

Figure 5.21: *Shown are two different surface profile plots (SVM (a) and (b)) obtained at the photodiode test sample. In both, (a) and (b), $V^2_{active} = V_{n-epi} = V_{n-sub} = 0\,V$, while in (a) $V^1_{active} = V_p = +0.3\,V$ and in (b) V^1_{active} was set to $+0.2\,V$. The second set of curves shows the resulting $V^2_{dynamic}$ as a function of the probe tip position. For SVM data (a) an SSRM measurement voltage of $V_{dc-sw} = 0.1\,V$ was applied, for the SVM data (b), V_{dc-sw} was set to $0.2\,V$. The effect of the different V_{dc-sw} values can be best observed at the n-hillock and n-epi position, below $400\,nm$.*

for the calculation of the individual $V_{dynamic}$ ($V_{dynamic}^1$ applied at p-shallow, $V_{dynamic}^2$ applied at n-epi). The resulting $V_{dynamic}^2$, as output by the SMUs, is shown for two voltages set by the SSRM control software (V_{dc-sw}). The equations for data calculation including values for the distinct contributions are included in figure 5.21.

5.4.3 Results

The experiments using the new SDVSRM technique, were conducted at the photodiode test sample. Every measurement delivers SVM and SSRM data, as both methods are combined in this two pass method.

Figure 5.22 a) shows the SSRM data obtained at $V_{dc-sw} = -50\,\text{mV}$. The devices active voltages were constant for V_{n-epi} and V_{p-sub}, both set to $0\,\text{V}$. V_p was varied during scanning as noted at the left hand side of the image. Starting from the bottom of the image at the devices equilibrium state, V_p was decreased in $1\,\text{V}$ steps, until a minimum V_p of $-6\,\text{V}$. The corresponding SVM and SSRM depth profiles are shown in 5.22 b) and c), respectively d) and e).

Both, SSRM image and profiles show no change in p-sub and n-epi carrier distribution during the experiment. This corresponds to the observed constant V_s with respect to changes in V_p and was expected due to the constant V_{active} voltages applied to both dopant areas.

Significant changes in effective carrier concentration could be observed within the first $4\,\mu\text{m}$ of the device. As expected, the moderately elevated n-hillock carrier concentration is decreasing for decreasing V_p. No resistivity minimum at the n-hillock center is observed for V_p values below $-2\,\text{V}$. This is in good agreement with CV-data and the SDSRM results. After this first depletion stage, a further decrease in shallow p voltage results in a significant stepwise increase in SCR width.

In contrast to all techniques discussed before, SDVSRM also reveals the topmost shallow p-layer (arrow with solid line in figure 5.22 d)). Due the possibility of its observation, the rather small increase in SCR width for the first stage of depletion, in contrast to the increase for $V_p < -2\,\text{V}$, can be visualized for the first time (arrows with dotted lines). The increase itself is the result of two effects: the increased reverse bias between the shallow p-layer and the adjacent n-layers and the decrease of the n-hillock carrier concentration.

This is the first evidence for the advantage of this new method, combining SVM and SSRM. The corresponding course of V_s in 5.22 c) and d) show a change in surface voltage, only for the first two microns, increasing in

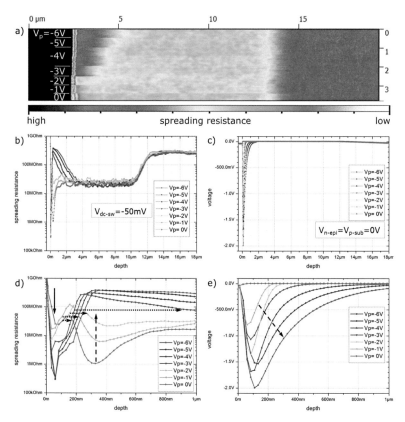

Figure 5.22: *SDVSRM results obtained at the photodiode test structure. In a) the resulting S5RM image using $V_{dc-sw} = -50\,\text{mV}$ is shown. The extracted profiles for each V_p setting are shown for the full depth in b) and for the first micrometer in d). The corresponding SVM data are shown in c) and e), respectively.*

145

depth and total voltage difference with increasing V_p. A significant difference between the applied V_p and the observed V_s at the shallow p depth is observed. This difference might be caused by a not optimal electrical contact. Furthermore, it shows the necessity of a combined SVM/SSRM approach for correct data acquisition.

The observed decrease in spreading resistance of the shallow p-layer with increasing V_p is not fully understood. The present deviation between expected and measured V_p might have an influence. On the other hand, comparable measurement results obtained in standard SSRM (figure 5.9) would suggest an opposite trend. Different laser light conditions can not be excluded as source of the deviations.

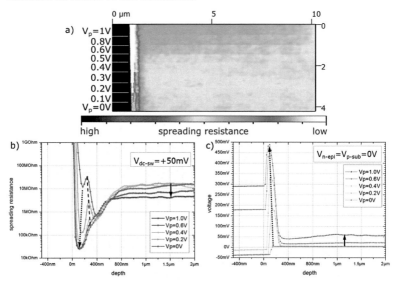

Figure 5.23: *SDVSRM results obtained at the photodiode test structure. In a) the resulting SSRM image using $V_{dc-sw} = +50$ mV is shown. The extracted profiles for each V_p setting are shown in b) and the corresponding SVM data are shown in c).*

For data acquisition of the local free carrier concentration in forward bias direction, a V_{dc-sw} of $+50$ mV was applied. As in the previous experiment, V_{n-epi} and V_{p-sub} were kept constant at 0 V. V_p was varied during scanning, starting from the devices equilibrium condition ($V_p = 0$ V) up to $+1$ V. The resulting SSRM image is shown in figure 5.23 a), the corresponding depth profiles in b) and the corresponding SVM data in c). Up to V_p values of +0.4 V only an increase in V_s at the shallow p layer is observed. As a result, local carrier concentration increases for this position. At the same time the width of the SCR between the shallow p and the adjacent n-layers decreases significantly. Although the observed maximum resistivity shifts slightly towards greater depths (dashed line), this might be a result of the decreased p-layer resistivity.

Above V_p values of +0.4 V the SCR disappears. At the same time an increase of the effective carrier concentration in the lowly doped n-epi and an increase in V_s for the n-epi is observed. This behavior can be attributed to the exceeding of the diodes built in potential and switching to forward bias conditions. In this state, the devices free carrier distribution shows significant differences for the deeper p-sub region and the p-n-junction between p-sub and n-epi, for the first time. An increase in V_s for the p-sub is observed.

In another experiment, the whole DUT potential was shifted with respect to the probe tip. Internal potential differences were kept constant. The SSRM data of the two cases is included in figure 5.23. The third stripe from top corresponds to $V_p = +1$ V, $V_{n-epi} = V_{p-sub} = 0$ V. The topmost stripe exhibits the same V_p potential difference with respect to V_{n-epi} and V_{p-sub}, but this time all voltages were decreased by 1 V. The resulting surface voltages V_s, shown in figure 5.24 a), clearly reveal the different voltage levels applied to the device. The corresponding SSRM data is shown in 5.24 b). The two profiles are in agreement, as expected for similar internal potential distribution. This experiment clearly demonstrates the major advantage of decoupling surface potential (V_s) and SSRM measurement voltage (V_{dc-sw}), as done by the new SDVSRM method.

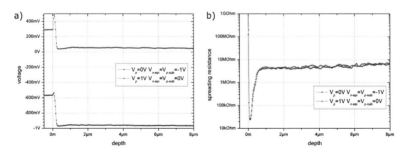

Figure 5.24: *Although the overall device potential is shifted with respect to the probe a), the observed SSRM data is unaffected b).*

5.4.4 Summary

A new SPM technique, combining SVM and SSRM in a two pass process, was conceived and applied successfully to an integrated device for the first time within this work. SDVSRM delivers complementary information, by the simultaneously mapping of surface voltage V_s and the correlating local free carrier concentration. Due to the implemented decoupling of the effective V_s and the SSRM measurement voltage (V_{dc-sw}), SDVSRM is an ideal candidate for two dimensional, high spatial resolution carrier concentration mapping of actively driven devices. Compared to the other methods applied to active device characterization, SDVSRM was the only one capable of imaging the shallow p-layer, SCR, n-hillock and n-epi at the same time in both, diode reverse and forward bias direction.

Furthermore, the possible deviations between V_s and V_{dc-sw} for devices under equilibrium conditions, as observed in SVM experiments described in the related section, could be circumvent by the application of SDVSRM.

6 Summary and Outlook

6.1 General Summary

Within this work a motivation for the employment of SSRM for characterization of dopant distributions in semiconductor devices was given. Alternative dopant profiling techniques, including their advantages and disadvantages were presented. Subsequently, the key aspects for successful SSRM measurement were discussed. For some of those aspects, e.g. the preparation methods available, improvements could be demonstrated. It was shown that the dependency of the obtained SSRM signal from the polarity of the applied dc voltage between sample and probe tip can be used to gain information on the present dopant type.

One focus of this work was the employment of the technique to different fields of application. SSRM proved to offer additional value in characterization of photovoltaic devices, demonstrated at the example of selective emitter analysis. Two analysis cases of integrated silicon devices emphasize the advantages of SSRM for failure analysis of dopant distribution related root cause. The methods unsurpassed combination of high lateral resolution, high dynamic range of the obtained spreading resistance signal and the ability to generate precisely located, specific site cross sections, free to choose in terms of position and orientation, are key for this kind of application. The large scan size available in SSRM was employed to characterize different power technologies with their associated large feature sizes. Moreover, by adaption of the measurement setup, fully processed power devices based on silicon carbide (SiC) could be analyzed.

The second focus of this work was dedicated to the characterization capabilities for actively driven devices. The wire bond approach within a cut in half package in combination with existing preparation methods delivered electrically non-disturbed cross sections for subsequent analysis. Several known SPM methods were applied at an actively driven test sample. Interestingly, even the standard SSRM results illustrate some of the aspects present within the active device. On the other hand, the significant differences between V_{dc-sw} and V_s limit the applicability of standard SSRM drastically. In SDSRM this limiting factor is tried to be compensated by the application of a lock-in amplifier. There are some interesting new aspects arising from this approach, e.g. the possibility to limit the sensitivity to a specific voltage source. Nevertheless, some of the observed phenomena could not be explained, making the interpretation of SDSRM results difficult. In contrast, SVM proved to be a very straight forward method, delivering complementary information on desired free carrier distribution.

A new method, combining SVM and SSRM in a two-pass process was introduced. SDVSRM offers the interpretability of classical SSRM, but ensures matching of the software set V_{dc-sw} and the devices local surface voltage, V_s. The local V_s dependent on V_{active} is obtained in the first pass. Based on V_{active}, the observed V_s and V_{dc-sw}, during SSRM data acquisition in the second pass, a dynamic voltage bias ($V_{dynamic}$) is applied at the individual device components, synchronized with the scanner position. The new method proved to be an ideal candidate for the characterization of actively driven devices.

An additional outcome of the SVM measurements was the evidence of significant deviations between V_{dc-sw} and V_s under certain circumstances, even in equilibrium conditions of the short circuited device in standard SSRM. Here again, SDVSRM has the potential to compensate the occurring differences in surface voltage V_s.

The main limitation of the SDVSRM method turned out to be the accurate determination of the local surface voltage. Occurring differences between measured and actual surface potential, inevitably lead to deviations of the obtained spreading resistance values from the true values.

6.2 Outlook

There is a lot of room for method development and further improvements. Today's methods for sample preparation offer either high quality cross section surfaces with respect to their electrical properties (cleaving) or high accuracy positioning and orientation for site specific preparation (grinding & polishing and FIB). A combination of both properties is highly desirable. Furthermore, a preparation technique reducing the required contact forces to a minimum could minimize tip- and sample wear out.

Probe resistances of $1\,k\Omega$ and above limit the sensitivity for high dopant concentrations. Development of low resistance probes would potentially improve SSRM analysis capability.

The comparability of SSRM results, obtained at different samples has to be significantly improved, in order to increase the possible applications of the method to the analysis of more subtle differences in dopant concentration, e.g. caused by process split experiments.

All experiments on actively driven devices were performed at the laterally homogenous photodiode sample. Although, the layer structure enabled the best possible investigation of the influence of changes in the applied V_{active}, future work must aim to apply the newly developed SDVSRM method to devices, exhibiting lateral differences in dopant concentration. The experienced deviations in local surface voltage from the expected values need further investigation. Advances in cross section preparation with respect to the presence of an amorphous damage layer and surface states should improve the situation. In order to gain additional information value, the obtained spreading resistance data could be calibrated.

Bibliography

[1] Ch. E. Fritts. A new form of selenium cell. *Am. J. Sci*, 26:465–472, 1883.

[2] J.C. Bose. Detector for electrical disturbances., March 29 1904. US Patent 755,840.

[3] J.E. Lilienfeld. Electric current control mechanism. *Canadian patent CA*, 272437(19):07, 1927.

[4] J.E. Lilienfeld. Method and apparatus for controlling electric currents, January 28 1930. US Patent 1,745,175.

[5] J. Bardeen and W.H. Brattain. The Transistor, A Semi-Conductor Triode. *Phys. Rev.*, 74:230–231, Jul 1948.

[6] J. Bardeen and W.H. Brattain. Physical Principles Involved in Transistor Action. *Phys. Rev.*, 75:1208–1225, Apr 1949.

[7] Michael Riordan. The Lost History of the Transistor. *IEEE Spectrum*, April 2004.

[8] G.E. Moore. Cramming More Components Onto Integrated Circuits. *Proceedings of the IEEE*, 86(1):82–85, Jan 1998.

[9] D. James. Intel Ivy Bridge unveiled - The first commercial tri-gate, high-k, metal-gate CPU. In *Custom Integrated Circuits Conference (CICC), 2012 IEEE*, pages 1–4, Sept 2012.

[10] J. Jang, H.-S. Kim, W. Cho, H. Cho, J. Kim, S.I. Shim, Y. Jang, J.-H. Jeong, B.-K. Son, D.W. Kim, Kihyun, J.-J. Shim, J.S. Lim, K.-H. Kim, S.Y. Yi, J.-Y. Lim, D. Chung, H.-C. Moon, S. Hwang, J.-W. Lee, Y.-H. Son, U.I. Chung, and W.-S. Lee. Vertical cell array using TCAT (Terabit Cell Array Transistor) technology for ultra high density NAND flash memory. In *VLSI Technology, 2009 Symposium on*, pages 192–193, June 2009.

153

[11] R.G. Mazur and D.H. Dickey. A Spreading Resistance Technique for Resistivity Measurements on Silicon. *Journal of The Electrochemical Society*, 113(3):255–259, 1966.

[12] T. Clarysse, W. Vandervorst, E.J.H. Collart, and A.J. Murrell. Electrical Characterization of Ultrashallow Dopant Profiles. *Journal of The Electrochemical Society*, 147(9):3569–3574, 2000.

[13] T. Clarysse, D. Vanhaeren, I. Hoflijk, and W. Vandervorst. Characterization of electrically active dopant profiles with the spreading resistance probe. *Materials Science and Engineering: R: Reports*, 47:123 – 206, 2004.

[14] R. Holm. *Electric Contacts Theory and Application*. Springer Verlag, New York, 1967.

[15] W.R. Thurber, R.L. Mattis, Y.M. Liu, and J.J. Filliben. Resistivity - Dopant Density Relationship for Phosphorus-Doped Silicon. *Journal of The Electrochemical Society*, 127(8):1807–1812, 1980.

[16] W. R. Thurber, R. L. Mattis, Y. M. Liu, and J. J. Filliben. Resistivity - dopant density relationship for boron - doped silicon. *Journal of The Electrochemical Society*, 127(10):2291–2294, 1980.

[17] T. Clarysse and W. Vandervorst. A new spreading resistance correction scheme combining variable radius and barrier resistance with epilayer matching. *Journal of Vacuum Science & Technology B*, 10(1):432–437, 1992.

[18] T. Clarysse, M. Caymax, P. De Wolf, T. Trenkler, W. Vandervorst, J. S. McMurray, J. Kim, C. C. Williams, J. G. Clark, and G. Neubauer. Epitaxial staircase structure for the calibration of electrical characterization techniques. *Journal of Vacuum Science & Technology B*, 16(1):394–400, 1998.

[19] P.A. Schumann and E.E. Gardner. Application of Multilayer Potential Distribution to Spreading Resistance Correction Factors. *Journal of The Electrochemical Society*, 116(1):87–91, 1969.

[20] S.M. Hu. Between carrier distributions and dopant atomic distribution in beveled silicon substrates. *Journal of Applied Physics*, 53(3):1499–1510, 1982.

[21] P.J. Severin. Measurement of resistivity of silicon by the spreading resistance method. *Solid-State Electronics*, 14(3):247 – 248, 1971.

[22] T. Clarysse and W. Vandervorst. Need to incorporate the real microcontact distribution in spreading resistance correction schemes. *Journal of Vacuum Science & Technology B*, 18(1):393–400, 2000.

[23] E.F. Gorey, C.P. Schneider, and M.R. Poponiak. Preparation and Evaluation of Spreading Resistance Probe Tip. *Journal of The Electrochemical Society*, 117(5):721–725, 1970.

[24] T. Clarysse, P. De Wolf, H. Bender, and W. Vandervorst. Recent insights into the physical modeling of the spreading resistance point contact. *Journal of Vacuum Science & Technology B*, 14(1):358–368, 1996.

[25] S. Minomura and H.G. Drickamer. Pressure induced phase transitions in silicon, germanium and some III-V compounds . *Journal of Physics and Chemistry of Solids*, 23(5):451 – 456, 1962.

[26] J.Z. Hu, L.D. Merkle, C.S. Menoni, and I.L. Spain. Crystal data for high-pressure phases of silicon. In *Phys. Rev. B* [24], pages 4679–4684.

[27] W. Vandervorst, T. Clarysse, J. Vanhellemont, and A. Romano-Rodriguez. Two-dimensional carrier profiling. *Journal of Vacuum Science & Technology B*, 10(1):449–455, 1992.

[28] V. Privitera, W. Vandervorst, and T. Clarysse. A Spreading Resistance-Based Technique for Two-Dimensional Carrier Profiling. *Journal of The Electrochemical Society*, 140(1):262–270, 1993.

[29] R. Castaing, B. Jouffrey, and G. Slodzian. Sur les possibilites danalyse locale dun echantillon par utilisation des son emission ionique secondaire. *Comptes rendus hebdomadaires des seances de l academie des sciences*, 251(8):1010–1012, 1960.

[30] H.J. Liebl and R.F.K. Herzog. Sputtering Ion Source for Solids. *Journal of Applied Physics*, 34(9):2893–2896, 1963.

[31] A. Benninghoven. Die Analyse monomolekularer Festkörperoberflächenschichten mit Hilfe der Sekundärionenemission. *Zeitschrift für Physik*, 230(5):403–417, 1970.

[32] E.J.H. Collart, K. Weemers, D.J. Gravesteijn, and J.G.M. van Berkum. Characterization of low-energy (100 eV-10 keV) boron ion implantation. *Journal of Vacuum Science & Technology B*, 16(1):280–285, 1998.

[33] M.G. Dowsett and G.A. Cooke. Two dimensional profiling using secondary ion mass spectrometry. *Journal of Vacuum Science & Technology B*, 10(1):353–357, 1992.

[34] V.A. Ukraintsev, P.J. Chen, J.T. Gray, C.F. Machala, L.K. Magel, and M.-C. Chang. High-resolution two-dimensional dopant characterization using secondary ion mass spectrometry. *Journal of Vacuum Science & Technology B*, 18(1):580–585, 2000.

[35] T. Ambridge and M.M. Faktor. An automatic carrier concentration profile plotter using an electrochemical technique. *Journal of Applied Electrochemistry*, 5(4):319–328, 1975.

[36] P. Blood. Capacitance-voltage profiling and the characterisation of III-V semiconductors using electrolyte barriers. *Semiconductor Science and Technology*, 1(1):7, 1986.

[37] Y. Akasaka, K. Horie, and S. Kawazu. Lateral spread of boron ions implanted in silicon. *Applied Physics Letters*, 21(4):128–129, 1972.

[38] C.P. Wu, E.C. Douglas, C.W. Mueller, and R. Williams. Techniques for lapping and staining ion-implanted layers. *Journal of The Electrochemical Society*, 126(11):1982–1988, 1979.

[39] R. Subrahmanyan. Methods for the measurement of two-dimensional doping profiles. *Journal of Vacuum Science & Technology B*, 10(1):358–368, 1992.

[40] J. Adrian, N. Rodriguez, F. Essely, G. Haller, C. Grosjean, A. Portavoce, and C. Girardeaux. Investigation of a new method for dopant characterization. *Microelectronics Reliability*, 47(9-11):1599 – 1603, 2007. 18th European Symposium on Reliability of Electron Devices, Failure Physics and Analysis.

[41] K.D. Yoo, C.D. Marsh, and G.R. Booker. Two-dimensional dopant concentration profiles from ultrashallow junction metal-oxide-semiconductor field-effect transistors using the etch/transmission electron microscopy method. *Applied Physics Letters*, 80(15):2687–2689, 2002.

[42] W.-T. Chang, T.-E. Hsieh, G. Zimmermann, and L. Wang. Advance static random access memory soft fail analysis using nanoprobing and junction delineation transmission electron microscopy. *Journal of Vacuum Science & Technology B*, 25(1):202–207, 2007.

[43] T.H.P. Chang and W.C. Nixon. Electron beam induced potential contrast on unbiased planar transistors. *Solid-State Electronics*, 10(7):701 – 704, 1967.

[44] R. Rosenkranz, S. Doering, W. Werner, L. Bartholomaus, and S. Eckl. Active Voltage Contrast for Failure Localization in Test Structures. In *International Symposium for Testing and Failure Analysis*, volume 32, page 316. ASM International; 1998, 2006.

[45] D.D. Perovic, M.R. Castell, A. Howie, C. Lavoie, T. Tiedje, and J.S.W. Cole. Field-emission SEM imaging of compositional and doping layer semiconductor superlattices. *Ultramicroscopy*, 58(1):104 – 113, 1995. Microscopy with Field Emission Electron Sources.

[46] I. Volotsenko, M. Molotskii, Z. Barkay, J. Marczewski, P. Grabiec, B. Jaroszewicz, G. Meshulam, E. Grunbaum, and Y. Rosenwaks. Secondary electron doping contrast: Theory based on scanning electron microscope and kelvin probe force microscopy measurements. *Journal of Applied Physics*, 107(1):–, 2010.

[47] A.K.W. Chee, R.F. Broom, C.J. Humphreys, and E.G.T. Bosch. A quantitative model for doping contrast in the scanning electron microscope using calculated potential distributions and Monte Carlo simulations. *Journal of Applied Physics*, 109(1):–, 2011.

[48] D. Venables, H. Jain, and D.C. Collins. Secondary electron imaging as a two-dimensional dopant profiling technique: Review and update. *Journal of Vacuum Science & Technology B*, 16(1):362–366, 1998.

[49] M. El-Gomati, F. Zaggout, H. Jayacody, S. Tear, and K. Wilson. Why is it possible to detect doped regions of semiconductors in low voltage SEM: a review and update. *Surface and Interface Analysis*, 37(11):901–911, 2005.

[50] P. Kazemian, A.C. Twitchett, C.J. Humphreys, and C. Rodenburg. Site-specific dopant profiling in a scanning electron microscope using focused ion beam prepared specimens. *Applied Physics Letters*, 88(21):–, 2006.

[51] G. Binnig and H. Rohrer. Scanning tunneling microscope, 08 1982.

[52] G. Binnig, H. Rohrer, Ch. Gerber, and E. Weibel. Tunneling through a controllable vacuum gap. *Applied Physics Letters*, 40(2):178–180, 1982.

[53] G. Binnig, H. Rohrer, Ch. Gerber, and E. Weibel. Surface studies by scanning tunneling microscopy. *Phys. Rev. Lett.*, 49:57–61, Jul 1982.

[54] G. Binnig, H. Rohrer, Ch. Gerber, and E. Weibel. 7 × 7 Reconstruction on Si(111) Resolved in Real Space. *Phys. Rev. Lett.*, 50:120–123, Jan 1983.

[55] G. Binnig and H. Rohrer. Scanning Tunneling Microscopy - from Birth to Adolescence (Nobel Lecture). *Angewandte Chemie International Edition in English*, 26(7):606–614, 1987.

[56] G. Binnig, C. Quate, and Ch. Gerber. Atomic Force Microscope. *Phys. Rev. Lett.*, 56:930–933, Mar 1986.

[57] Y. Martin, C.C. Williams, and H.K. Wickramasinghe. Atomic force microscope - force mapping and profiling on a sub 100åscale. *Journal of Applied Physics*, 61(10):4723–4729, 1987.

[58] G. Meyer and N.M. Amer. Novel optical approach to atomic force microscopy. *Applied Physics Letters*, 53(12):1045–1047, 1988.

[59] F.J. Giessibl. Atomic Resolution of the Silicon (111)-(7x7) Surface by Atomic Force Microscopy. *Science*, 267(5194):68–71, 1995.

[60] J.R. Matey and J. Blanc. Scanning capacitance microscopy. *Journal of Applied Physics*, 57(5):1437–1444, 1985.

[61] J.A. Slinkman, H.K. Wickramasinghe, and C.C. Williams. Scanning capacitance - voltage microscopy, November 12 1991. US Patent 5,065,103.

[62] E. Bussmann and C.C. Williams. Sub-10 nm lateral spatial resolution in scanning capacitance microscopy achieved with solid platinum probes. *Review of Scientific Instruments*, 75(2):422–425, 2004.

[63] V. Raineri and S. Lombardo. Effective channel length and base width measurements by scanning capacitance microscopy. *Journal of Vacuum Science & Technology B*, 18(1):545–548, 2000.

[64] R. Stephenson, P. De Wolf, T. Trenkler, T. Hantschel, T. Clarysse, P. Jansen, and W. Vandervorst. Practicalities and limitations of scanning capacitance microscopy for routine integrated circuit characterization. *Journal of Vacuum Science & Technology B*, 18(1):555–559, 2000.

[65] J. Isenbart, A. Born, and R. Wiesendanger. The physical principles of scanning capacitance spectroscopy. *Applied Physics A*, 72(2):S243–S251, 2001.

[66] D. Goghero, V. Raineri, and F. Giannazzo. Study of interface states and oxide quality to avoid contrast reversal in scanning capacitance microscopy. *Applied Physics Letters*, 81(10):1824–1826, 2002.

[67] K.M. Wong, W.K. Chim, and J. Yan. Physical mechanism of oxide interfacial traps, carrier mobility degradation and series resistance on contrast reversal in scanning-capacitance-microscopy dopant concentration extraction. *Applied Physics Letters*, 87(5):–, 2005.

[68] H. Lichte and M. Lehmann. Electron holography—basics and applications. *Reports on Progress in Physics*, 71(1):016102, 2008.

[69] T.F. Kelly and M.K. Miller. Atom probe tomography. *Review of Scientific Instruments*, 78(3):–, 2007.

[70] J. Boussey. Stripping Hall effect, sheet and spreading resistance techniques for electrical evaluation of implanted silicon layers. *Microelectronic Engineering*, 40(3–4):275 – 284, 1998. Electrical and Physical Characterization.

[71] C. Shafai, D.J. Thomson, and M. Simard-Normandin. Two-dimensional delineation of semiconductor doping by scanning resistance microscopy. *Journal of Vacuum Science & Technology B*, 12(1):378–382, 1994.

[72] J.N. Nxumalo, D.T. Shimizu, and D.J. Thomson. Cross-sectional imaging of semiconductor device structures by scanning resistance microscopy. *Journal of Vacuum Science & Technology B*, 14(1):386–389, 1996.

[73] J.N. Nxumalo, D.T. Shimizu, D.J. Thomson, and M. Simard-Normadin. High-resolution cross-sectional imaging of MOSFETs by scanning resistance microscopy. *Electron Device Letters, IEEE*, 18(2):71–73, Feb 1997.

[74] W. Vandervorst and M. Meuris. Method and apparatus for resistance measurements on a semiconductor element, 1992.

[75] P. De Wolf, J. Snauwaert, L. Hellemans, T. Clarysse, W. Vandervorst, M. D'Olieslaeger, and D. Quaeyhaegens. Lateral and vertical dopant profiling in semiconductors by atomic force microscopy using conducting tips. *Journal of Vacuum Science & Technology A*, 13(3):1699–1704, 1995.

[76] P. De Wolf, M. Geva, T. Hantschel, W. Vandervorst, and R.B. Bylsma. Two-dimensional carrier profiling of InP structures using scanning spreading resistance microscopy. *Applied Physics Letters*, 73(15):2155–2157, 1998.

[77] P. De Wolf, R. Stephenson, S. Biesemans, P. Jansen, G. Badenes, K. De Meyer, and W. Vandervorst. Direct measurement of L-eff and channel profile in MOSFETs using 2-D carrier profiling techniques. *International Electron Devices Meeting 1998 - Technical Digest*, pages 559–562, 1998.

[78] P. Eyben. Probing semiconductor technology and devices with scanning spreading resistance microscopy. In Sergei Kalinin and Alexei Gruverman, editors, *Scanning Probe Microscopy*, pages 31–87–. Springer New York, 2007.

[79] P. Eyben, S.-C. Vemula, T. Noda, and W. Vandervorst. Two-dimensional carrier profiling with sub-nm resolution using SSRM: From basic concept to TCAD calibration and device tuning. pages 74–78, 2009.

[80] L. Zhang, H. Tanimoto, K. Adachi, and A. Nishiyama. 1-nm spatial resolution in carrier profiling of ultrashallow junctions by scanning spreading resistance microscopy. *Ieee Electron Device Letters*, 29(7):799–801, 2008.

[81] P. Eyben, D. Degryse, and W. Vandervorst. On the spatial resolution of scanning spreading resistance microscopy : experimental assessment and electromechanical modeling. *Characterization and Metrology for ULSI Technology 2005*, 788:264–269, 2005.

[82] P. Eyben, F. Clemente, K. Vanstreels, G. Pourtois, T. Clarysse, E. Duriau, T. Hantschel, K. Sankaran, J. Mody, W. Vandervorst,

K. Mylvaganam, and L. Zhang. Analysis and modeling of the high vacuum scanning spreading resistance microscopy nanocontact on silicon. *Journal of Vacuum Science & Technology B: Microelectronics and Nanometer Structures*, 28(2):401–406, 2010.

[83] K. Mylvaganam, L.C. Zhang, P. Eyben, J. Mody, and W. Vandervorst. Evolution of metastable phases in silicon during nanoindentation: mechanism analysis and experimental verification. *Nanotechnology*, 20(30):305705, 2009.

[84] T. Hantschel, P. Niedermann, T. Trenkler, and W. Vandervorst. Highly conductive diamond probes for scanning spreading resistance microscopy. *Applied Physics Letters*, 76(12):1603–1605, 2000.

[85] N. Yasutake, A. Nomachi, H. Itokawa, T. Morooka, L. Zhang, T. Fukushima, H. Harakawa, I. Mizushima, A. Azuma, and Y. Toyosihma. Comprehensive study of S/D engineering for 32 nm node CMOS in direct silicon bonded (DSB) technology. *Solid-State Electronics*, 53(7):694–700, July 2009.

[86] P. Eyben, Ja. Clarysse, T.and Mody, A. Nazir, A. Schulze, T. Hantschel, and W. Vandervorst. Two-dimensional carrier mapping at the nanometer-scale on 32nm node targeted p-MOSFETs using high vacuum scanning spreading resistance microscopy. *Solid-State Electronics*, 71(0):69–73, May 2012.

[87] P. Eyben, D. Alvarez, M. Jurczak, R. Rooyackers, A. De Keersgieter, E. Augendre, and W. Vandervorst. Analysis of the two-dimensional-dopant profile in a 90 nm complementary metal-oxide-semiconductor technology using scanning spreading resistance microscopy. *Journal of Vacuum Science & Technology B*, 22(1):364–368, 2004.

[88] T. Schweinbock, S. Schomann, D. Alvarez, M. Buzzo, W. Frammelsberger, P. Breitschopf, and G. Benstetter. New trends in the application of scanning probe techniques in failure analysis. In *Microelectronics Reliability*, volume 44, pages 1541–1546, OCT 04-08 2004.

[89] L. Zhang. Site-specific and high-spatial-resolution scanning spreading resistance microscopy (SSRM) and its applications to Si devices. In *Junction Technology (IWJT), 2012 12th International Workshop on*, pages 89–93, 2012.

[90] J. Mody, P. Eyben, E. Augendre, O. Richard, and W. Vandervorst. Toward extending the capabilities of scanning spreading resistance microscopy for fin field-effect-transistor-based structures. *Journal of Vacuum Science & Technology B*, 26(1):351–356, 2008.

[91] J. Mody, P. Eyben, W. Polspoel, M. Jurczak, and W. Vandervorst. Scanning Spreading Resistance Microscopy for 3D-Carrier Profiling in FinFET-Based Structures. *Doping Engineering For Front-end Processing*, 1070:49–55, 2008.

[92] P. De Wolf, T. Clarysse, and W. Vandervorst. Quantification of nanospreading resistance profiling data. *Journal of Vacuum Science & Technology B*, 16(1):320–326, 1998.

[93] P. De Wolf, J. Snauwaert, T. Clarysse, W. Vandervorst, and L. Hellemans. Characterization of a point-contact on silicon using force microscopy-supported resistance measurements. *Applied Physics Letters*, 66(12):1530–1532, 1995.

[94] M.C. Gupta and A.L. Ruoff. Static compression of silicon in the [100] and in the [111] directions. *Journal of Applied Physics*, 51(2):1072–1075, 1980.

[95] D.E. Kim and S.I. Oh. Deformation pathway to high-pressure phases of silicon during nanoindentation. *Journal of Applied Physics*, 104(1):–, 2008.

[96] J.J. Marchand and V.K. Truong. Quantitative model for current-voltage characteristics of metal point contacts on silicon rectifying junctions. *Journal of Applied Physics*, 54(12):7034–7040, 1983.

[97] P. Eyben, S. Denis, T. Clarysse, and W. Vandervorst. Progress towards a physical contact model for scanning spreading resistance microscopy. *Materials Science and Engineering B-solid State Materials For Advanced Technology*, 102(1-3):132–137, 2003.

[98] J. Lutz, H. Schlangenotto, U. Scheuermann, and R.D. Doncker. Semiconductor power device. *Heidelberg: Springer*, 201(1):327, 2011.

[99] W. Vandervorst, J.L. Everaert, E. Rosseel, M. Jurczak, T. Hoffman, P. Eyben, J. Mody, G. Zschaetzsch, S. Koelling, M. Gilbert, T. Poon, J. del Eyben, M. Foad, R. Duffy, and B.J. Pawlak. Conformal Doping

of FINFETs: a Fabrication and Metrology Challenge. *Ion Implantation Technology 2008*, 1066:449–456, 2008.

[100] L. Zhang and I. Zarudi. Towards a deeper understanding of plastic deformation in mono-crystalline silicon. *International Journal of Mechanical Sciences*, 43(9):1985 – 1996, 2001. Selected Papers from AEPA 2000.

[101] U. Muehle, J. Steinhoff, and L. Hillmann. Influence of FIB-acceleration Voltage on Lateral Damage of Silicon based TEM samples. *Microscopy and Microanalysis*, 13:102–103, 9 2007.

[102] S. Beuer, V. Yanev, M. Rommel, A.J. Bauer, and H. Ryssel. SSRM characterisation of FIB induced damage in silicon. *Journal of Physics: Conference Series*, 100(5):052007–, 2008.

[103] G. Spoldi, S. Beuer, M. Rommel, V. Yanev, A.J. Bauer, and H. Ryssel. Experimental observation of FIB induced lateral damage on silicon samples. *Microelectronic Engineering*, 86(4-6):548–551, April 2009.

[104] M. Rommel, G. Spoldi, V. Yanev, S. Beuer, B. Amon, J. Jambreck, S. Petersen, A. J. Bauer, and L. Frey. Comprehensive study of focused ion beam induced lateral damage in silicon by scanning probe microscopy techniques. *J. Vac. Sci. Technol. B*, 28(3):595–607, May 2010.

[105] M. Mitome. Ultrathin specimen preparation by a low-energy Ar-ion milling method. *Microscopy*, 62(2):321–326, 2013.

[106] R.M. Langford and A.K. Petford-Long. Preparation of transmission electron microscopy cross-section specimens using focused ion beam milling. *Journal of Vacuum Science & Technology A*, 19(5):2186–2193, 2001.

[107] T. Hantschel, R. Stephenson, T. Trenkler, P. De Wolf, and W. Vandervorst. Characterization of silicon cantilevers with integrated pyramidal metal tips in atomic force microscopy. *Design, Test, and Microfabrication of Mems and Moems, Pts 1 and 2*, 3680:994–1005, 1999.

[108] T. Hantschel, C. Demeulemeester, P. Eyben, V. Schulz, O. Richard, H. Bender, and W. Vandervorst. Conductive diamond tips with sub-nanometer electrical resolution for characterization of nanoelectronics device structures. *physica status solidi (a)*, 206(9):2077–2081, 2009.

[109] S. Doering, U. Zimmermann, P. Rodger, and W. Werner. Analysis of deep trench node leakage depth by applying current imaging technique at silicon level. In *ISTFA 2008 Proceedings*, 2008.

[110] T. Hantschel, S. Slesazeck, P. Niedermann, P. Eyben, and W. Vandervorst. Integrating diamond pyramids into metal cantilevers and using them as electrical AFM probes. *Microelectronic Engineering*, 57-8:749–754, 2001.

[111] P. Eyben, M. Xu, N. Duhayon, T. Clarysse, S. Callewaert, and W. Vandervorst. Scanning spreading resistance microscopy and spectroscopy for routine and quantitative two-dimensional carrier profiling. *Journal of Vacuum Science & Technology B*, 20(1):471–478, 2002.

[112] D.M. Chapin, C.S. Fuller, and G.L. Pearson. A New Silicon p-n Junction Photocell for Converting Solar Radiation into Electrical Power. *Journal of Applied Physics*, 25(5):676–677, 1954.

[113] P. Eyben, F. Seidel, T. Hantschel, A. Schulze, A. Lorenz, A.U. De Castro, D. Van Gestel, J. John, J. Horzel, and W. Vandervorst. Development and optimization of scanning spreading resistance microscopy for measuring the two-dimensional carrier profile in solar cell structures. *physica status solidi (a)*, 208(3):596–599, 2011.

[114] J. Mandelkorn and J.H. Lamneck Jr. Simplified fabrication of back surface electric field silicon cells and novel characteristics of such cells. *Solar Cells*, 29(2):121 – 130, 1990.

[115] J.G. Fossum. Physical operation of back-surface-field silicon solar cells. *Electron Devices, IEEE Transactions on*, 24(4):322–325, Apr 1977.

[116] E. Van Kerschaver and G. Beaucarne. Back-contact solar cells: a review. *Progress in Photovoltaics: Research and Applications*, 14(2):107–123, 2006.

[117] G. Hahn. Status of selective emitter technology. In *25th European Photovoltaic Solar Energy Conference and Exhibition / 5th World Conference on Photovoltaic Energy Conversion*, 2010.

[118] M.A. Green and P. Campbell. Light trapping properties of pyramidally textured surfaces. *Journal of Applied Physics*, 62(1):243–249, July 1987.

[119] Boehme. Laser processes for high efficiency cell concepts. In *ISL 2010 - International Symposium on Laser-Micromachining*, 2010.

[120] S. Doering, S. Jakschick, and T. Mikolajick. Scanning Spreading Resistance Microscopy as technique for silicon solar cell emitter structure characterization. In *Proc. 26th Eur. Photovoltaic Solar Energy Conf. Exh*, pages 122–125, 2011.

[121] P. Ferrada, R. Harney, E. Wefringhaus, S. Doering, S. Jakschick, T. Mikolajick, P. Eyben, T. Hantschel, W. Vandervorst, M. Weiss, and J. Lossen. Local doping profiles for height-selective emitters determined by Scanning Spreading Resistance Microscopy (SSRM). *Photovoltaics, IEEE Journal of*, 3(1):168–174, Jan 2013.

[122] S. Doering, A. Wachowiak, U. Winkler, M. Richter, J.Goehler, H. Roetz, S. Eckl, and T. Mikolajick. Scanning Spreading Resistance Microscopy analysis of locally blocked implant sites. *Microelectronic Engineering*, 122(0):77–81, 2014.

[123] S. Doering, A. Wachowiak, M. Rochel, Ch. Nowak, M. Hoffmann, U. Winkler, M. Richter, H. Roetz, S. Eckl, and T. Mikolajick. Polycrystalline silicon gate originated CMOS device failure investigated by Scanning Spreading Resistance Microscopy. *Microelectronic Engineering*, 142:40 – 46, 2015.

[124] C.Y. Wong, J.Y. Sun, Y. Taur, C. S. Oh, R. Angelucci, and B. Davari. Doping of n^+ and p^+ polysilicon in a dual-gate CMOS process. In *Electron Devices Meeting, 1988. IEDM '88. Technical Digest., International*, pages 238–241, 1988.

[125] N.C.-C. Lu, J.M. Sung, H.C. Kirsch, S.J. Hillenius, T. E. Smith, and L. Manchanda. Anomalous C-V characteristics of implanted poly MOS structure in n^+/p^+ dual-gate CMOS technology. *Electron Device Letters, IEEE*, 10(5):192–194, 1989.

[126] S.-W. Lee, C. Liang, C.-S. Pan, W. Lin, and J.B. Mark. A study on the physical mechanism in the recovery of gate capacitance to C_{ox} in implanted polysilicon MOS structures. *Electron Device Letters, IEEE*, 13(1):2–4, 1992.

[127] W.W. Lin and C. Liang. Separation of d.c. and a.c. competing effect of polysilicon-gate depletion in deep submicron CMOS circuit

performance. *Solid-State Electronics*, 39(9):1391–1393, September 1996.

[128] H.P. Tuinhout, A.H. Montree, J. Schmitzb, and P.A. Stolk. Effects of gate depletion and boron penetration on matching of deep submicron CMOS transistors. In *Electron Devices Meeting, 1997. IEDM '97. Technical Digest., International*, pages 631–634, 1997.

[129] T. Kamins. *Polycrystalline silicon for integrated circuits and displays (2nd ed.)*. Kluwer Academic Publishers, 1998.

[130] H. Puchner and S. Selberherr. An advanced model for dopant diffusion in polysilicon. *Electron Devices, IEEE Transactions on*, 42(10):1750–1755, 1995.

[131] S. Stokowski and M. Vaez-Iravani. Wafer inspection technology challenges for ULSI manufacturing. *AIP Conference Proceedings*, 449(1):405–415, 1998.

[132] R. Rosenkranz. Failure localization with active and passive voltage contrast in FIB and SEM. *Journal of Materials Science: Materials in Electronics*, 22(10):1523–1535, 2011.

[133] R. Mulder, S. Subramanian, and T. Chrastecky. Case Studies in Atomic Force Probe Analysis. In *ISTFA Proceedings*, 2006.

[134] P. Egger, S. Mueller, and M. Stiftinger. A New Approach for SRAM Soft Defect Root Cause Identification. In *ISTFA Proceedings*, 2007.

[135] S. Doering, R. Rudolf, M. Pinkert, H. Roetz, C. Wagner, S. Eckl, M. Strasser, A. Wachowiak, and T. Mikolajick. Scanning Spreading Resistance Microscopy for failure analysis of nLDMOS devices with decreased breakdown voltage. *Microelectronics Reliability*, 54(9–10):2128 – 2132, 2014. ESREF 2014.

[136] X.-B. Chen and C. Hu. Optimum doping profile of power MOSFET epitaxial layer. *Electron Devices, IEEE Transactions on*, 29(6):985–987, Jun 1982.

[137] G. Deboy, N. Marz, J.-P. Stengl, H. Strack, J. Tihanyi, and H. Weber. A new generation of high voltage MOSFETs breaks the limit line of silicon. In *Electron Devices Meeting, 1998. IEDM '98. Technical Digest., International*, pages 683–685, Dec 1998.

[138] S. Ono, L. Zhang, H. Ohta, M. Watanabe, W. Saito, S. Sato, H. Sugaya, and M. Yamaguchi. Development of 600V-class trench filling SJ-MOSFET with SSRM analysis technology. In *Power Semiconductor Devices & IC's, 2009. ISPSD 2009. 21st International Symposium on*, pages 303–306, 2009.

[139] L. Zhang, M. Koike, M. Ono, S. Itai, K. Matsuzawa, S. Ono, W. Saito, M. Yamaguchi, Y. Hayase, and K. Hara. Comprehensive 2D-carrier profiling of low-doping region by high-sensitivity scanning spreading resistance microscopy (SSRM) for power device applications. *Microelectronics Reliability*, pages –, 2015.

[140] J.J. Berzelius. Untersuchungen über die Flussspathsäure und deren merkwürdigsten Verbindungen. *Annalen der Physik*, 77(6):169–230, 1824.

[141] A.G. Acheson. Characterization of Electrical Properties of 4H-SiC by Imaging Techniques. *Engl. Pat*, 17911, 1892.

[142] R. Rupp, T. Laska, O. Haberlen, and M. Treu. Application specific trade-offs for WBG SiC, GaN and high end Si power switch technologies. In *Electron Devices Meeting (IEDM), 2014 IEEE International*, pages 2.3.1–2.3.4, Dec 2014.

[143] R. Rupp, R. Gerlach, A. Kabakow, R. Schorner, C. Hecht, R. Elpelt, and M. Draghici. Avalanche behaviour and its temperature dependence of commercial SiC MPS diodes: Influence of design and voltage class. In *Power Semiconductor Devices IC's (ISPSD), 2014 IEEE 26th International Symposium on*, pages 67–70, June 2014.

[144] P. Friedrichs. SiC Power Devices as Enabler for High Power Density - Aspects and Prospects. In *Materials Science Forum*, volume 778-780, pages 1104–1109, 2014.

[145] A. Kumar and M.S. Aspalli. SiC: An advanced semiconductor material for power devices. *International Journal of Research in Engineering and Technology*, (03):248–252, 2014.

[146] Y. Tanaka, K. Yano, M. Okamoto, A. Takatsuka, K. Fukuda, M. Kasuga, K. Arai, and T. Yatsuo. Fabrication of 700V SiC-SIT with Ultra-Low On-Resistance of $1.01 m\Omega cm^2$. In *Silicon Carbide and Related Materials 2005*, volume 527-529, pages 1219–1222, 2005.

[147] J. Osterman, L. Abtin, U. Zimmermann, M.S. Janson, S. Anand, C. Hallin, and A. Hallen. Scanning spreading resistance microscopy of aluminum implanted 4H-SiC. *Materials Science and Engineering: B*, 102(1-3):128 – 131, 2003. E-MRS 2002 Symposium E: Advanced Characterisation of Semiconductors.

[148] A. Suchodolskis, A. Hallén, M.K. Linnarsson, J. Österman, and U.O. Karlsson. Ion implantation damage annealing in 4H-SiC monitored by scanning spreading resistance microscopy. *Thin Solid Films*, 515(2):611 – 614, 2006.

[149] M. Yoshida, A. Onodera, M. Ueno, K. Takemura, and O. Shimomura. Pressure-induced phase transition in SiC. *Phys. Rev. B*, 48:10587–10590, Oct 1993.

[150] M. Usman, M. Nawaz, and A. Hallen. Position-Dependent Bulk Traps and Carrier Compensation in 4H-SiC Bipolar Junction Transistors. *Electron Devices, IEEE Transactions on*, 60(1):178–185, 2013.

[151] A. Hallen, A. Suchodolskis, J. Osterman, L. Abtin, and M. Linnarsson. Annealing of Al implanted 4H silicon carbide. *Physica Scripta*, T126:37–40, 2006.

[152] Z.K. Lee, M.B. McIlrath, and D.A. Antoniadis. Two-dimensional doping profile characterization of MOSFETs by inverse modeling using I-V characteristics in the subthreshold region. *Electron Devices, IEEE Transactions on*, 46(8):1640–1649, Aug 1999.

[153] T. Trenkler, P. De Wolf, J. Snauwaert, Z. Qamhieh, W. Vandervorst, and L. Hellemans. Local potential measurements in silicon devices using atomic force microscopy with conductive tips. In *Solid State Device Research Conference, 1995. ESSDERC '95. Proceedings of the 25th European*, pages 477–481, Sept 1995.

[154] L.C. Hellemans, T. Trenkler, P. de Wolf, and W. Vandervorst. Method for measuring the electrical potential in a semiconductor element, March 3 1998. US Patent 5,723,981.

[155] T. Trenkler, P. De Wolf, W. Vandervorst, and L. Hellemans. Nanopotentiometry: Local potential measurements in complementary metal–oxide–semiconductor transistors using atomic force microscopy. *J. Vac. Sci. Technol. B*, 16(1):367–372, January 1998.

[156] T. Trenkler, R. Stephenson, P. Jansen, W. Vandervorst, and L. Helle-mans. New aspects of nanopotentiometry for complementary metal–oxide–semiconductor transistors. *J. Vac. Sci. Technol. B*, 18(1):586–594, January 2000.

[157] V.V. Zavyalov, J.S. McMurray, S.D. Stirling, C.C. Williams, and H. Smith. Two dimensional dopant and carrier profiles obtained by scanning capacitance microscopy on an actively biased cross-sectioned metal–oxide–semiconductor field-effect transistor. In *J. Vac. Sci. Technol. B*, volume 18, pages 549–554, 2000.

[158] C.Y. Nakakura, P. Tangyunyong, D.L. Hetherington, and M.R. Shaneyfelt. Method for the study of semiconductor device operation using scanning capacitance microscopy. *Rev. Sci. Instrum.*, 74(1):127–133, January 2003.

[159] D. Ban, E.H. Sargent, St.J. Dixon-Warren, T. Grevatt, G. Knight, G. Pakulski, A.J. SpringThorpe, R. Streater, and J.K. White. Two-dimensional profiling of carriers in a buried heterostructure multi-quantum-well laser: Calibrated scanning spreading resistance microscopy and scanning capacitance microscopy. *Journal of Vacuum Science & Technology B*, 20(5):2126–2132, 2002.

[160] D. Ban, E.H. Sargent, St.J. Dixon-Warren, I. Calder, A.J. SpringThorpe, R. Dworschak, G. Este, and J.K. White. Direct imaging of the depletion region of an InP p–n junction under bias using scanning voltage microscopy. *Applied Physics Letters*, 81(26):5057–5059, 2002.

[161] D. Ban, E.H. Sargent, St.J. Dixon-Warren, I. Calder, T. Grevatt, G. Knight, and J.K. White. Two-dimensional transverse cross-section nanopotentiometry of actively driven buried-heterostructure multiple-quantum-well lasers. *Journal of Vacuum Science & Technology B: Microelectronics and Nanometer Structures*, 20(6):2401–2407, 2002.

[162] D. Ban, E.H. Sargent, St.J. Dixon-Warren, K. Hinzer, J.K. White, and A.J. SpringThorpe. Scanning voltage microscopy on active semi-conductor lasers: the impact of doping profile near an epitaxial growth interface on series resistance. *Quantum Electronics, IEEE Journal of DOI - 10.1109/JQE.2004.828262*, 40(6):651–655, 2004.

[163] D. Ban, E.H. Sargent, St.J. Dixon-Warren, G. Letal, K. Hinzer, J.K. White, and D.G. Knight. Scanning voltage microscopy on buried

heterostructure multiquantum-well lasers: identification of a diode current leakage path. *Quantum Electronics, IEEE Journal of DOI - 10.1109/JQE.2003.821539*, 40(2):118–122, 2004.

[164] D. Ban, E.H. Sargent, and St.J. Dixon-Warren. Scanning differential spreading resistance microscopy on actively driven buried heterostructure multiquantum-well lasers. *Quantum Electronics, IEEE Journal of DOI - 10.1109/JQE.2004.830174*, 40(7):865–870, 2004.

[165] D. Ban, E.H. Sargent, and St. J Dixon-Warren. Direct observation of electron overbarrier leakage in actively driven buried heterostructure multi-quantum-well lasers. volume 5577, pages 66–73, 2004.

[166] S.B. Kuntze, D. Ban, E.H. Sargent, J.K. Dixon-Warren, St.J.and White, and K. Hinzer. Scanning Voltage Microscopy. In Sergei Kalinin and Alexei Gruverman, editors, *Scanning Probe Microscopy*, pages 561–600. Springer New York, 2007.

[167] R.A. Oliver. Advances in AFM for the electrical characterization of semiconductors. *Reports on Progress in Physics*, 71(7):076501, 2008.

[168] A. Wachowiak, S. Slesazeck, P. Jordan, J. Holz, and T. Mikolajick. New color sensor concept based on single spectral tunable photodiode. In *Solid-State Device Research Conference (ESSDERC), 2013 Proceedings of the European*, pages 127–130, Sept 2013.

Scientific contributions

- **S. Doering**, S. Jakschick, and T. Mikolajick. Scanning Spreading Resistance Microscopy as technique for silicon solar cell emitter structure characterization. In *Proc. 26th Eur. Photovoltaic Solar Energy Conf. Exh*, pages 122–125, 2011.

- P. Ferrada, R. Harney, E. Wefringhaus, **S. Doering**, S. Jakschick, T. Mikolajick, P. Eyben, T. Hantschel, W. Vandervorst, M. Weiss, and J. Lossen. Local doping profiles for height-selective emitters determined by Scanning Spreading Resistance Microscopy (SSRM). *Photovoltaics, IEEE Journal of*, 3(1):168–174, Jan 2013.

- **S. Doering**, A. Wachowiak, U. Winkler, M. Richter, J.Goehler, H. Roetz, S. Eckl, and T. Mikolajick. Scanning Spreading Resistance Microscopy analysis of locally blocked implant sites. *Microelectronic Engineering*, 122(0):77–81, 2014.

- **S. Doering**, R. Rudolf, M. Pinkert, H. Roetz, C. Wagner, S. Eckl, M. Strasser, A. Wachowiak, and T. Mikolajick. Scanning Spreading Resistance Microscopy for failure analysis of nLDMOS devices with decreased breakdown voltage. *Microelectronics Reliability*, 54(9–10):2128 – 2132, 2014. ESREF 2014.

- **S. Doering**, A. Wachowiak, M. Rochel, Ch. Nowak, M. Hoffmann, U. Winkler, M. Richter, H. Roetz, S. Eckl, and T. Mikolajick. Polycrystalline silicon gate originated CMOS device failure investigated by Scanning Spreading Resistance Microscopy. *Microelectronic Engineering*, 142:40 – 46, 2015.

Bisher erschienene Bände der Schriftenreihe Research at NaMLab

Herausgeber: Thomas Mikolajick   ISSN 2191-7167

1   Thomas Melde   Modellierung und Charakterisierung des elektrischen
                        Verhaltens von haftstellen-basierten Flash-Speicherzellen
                        ISBN 978-3-8325-2748-8   41.50 €

2   Guntrade Roll   Leakage Current and Defect Characterization of Short
                        Channel MOSFETs
                        ISBN 978-3-8325-3261-1   49.50 €

3   Andreas Krause   Ultrathin Calcium Titanate Capacitors: Physics and
                        Application
                        ISBN 978-3-8325-3724-1   35.00 €

4   Ekaterina Yurchuk   Electrical Characterisation of Ferroelectric Field Effect
                        Transistors based on Ferroelectric HfO_2 Thin Films
                        ISBN 978-3-8325-4003-6   40.50 €

Alle erschienenen Bücher können unter der angegebenen ISBN direkt online
(http://www.logos-verlag.de) oder per Fax (030 - 42 85 10 92)
beim Logos Verlag Berlin bestellt werden.